The Impossibility

of

Extraterrestrial Life

by Ben Tripp M. A. Sc., P. Eng.

.

The Impossibility of Extraterrestrial Life

By Ben Tripp M. A. Sc., P. Eng.
Illustrations by Carolyn Tripp B.A.

Other books by the same author:
1. The Window of Life
2. Fairytales for Adults
3. Concerning the Birth of Christ
4. The Asteroid Theory of the Flood and the Ice Age
5. The non-Myths of the Bible
6. Elements of Providence during the Genesis Flood
7. Too Much Carbon
8. Climate Change and Holy Writ
9. Time is running Out
   (more info at http://benatripp.wix.com/window-of-life)

About the author:

Ben earned Bachelors and Masters of Applied Science degrees in Engineering from the University of Waterloo, Ontario, and has worked as a consulting engineer on such projects as controls for large telescopes and test equipment for the CanadArm. He holds patents for innovations involving the recycling of used tires into fence boards and a novel ground coil arrangement for geothermal heat pumps.

Ben's interest in the current topic, and his related background reading and research span several decades and have culminated in what he purports to be a credible, cohesive and insightful discussion of one aspect of Nature. It is his hope that these observations and opinions will be helpful to many in their own investigations.

The Impossibility of Extraterrestrial Life

1st printing 2021

A bibliography and list of references has been included in the appendix. Numbers embedded in the text (e. g. (45)) are reference numbers.

ISBN   978-1-7751150-3-8 Soft cover version
ISBN   978-0-7751150-4-5 Electronic version

This work is dedicated to my family

To my dear wife:
Judith Anne
The love of my life

To my children:
Bryan, Rebecca, Daniel and Carolyn
The great blessing of my life

To my dear grandchildren:
Evelyn, Ayla, Zoe, Izzy and Ben

And to my bonus child:
Andrea

May this humble epistle assist them in their search for truth.

Table of Contents

Illustrations

1.0 Foreword

The idea that there is life elsewhere in the universe is very popular and does have some rationale to back it up. The most common argument recognizes that there are a very great number of stars in the milky-way galaxy and with such a large number there surely must be places where life does exist. This idea is not new and has been around for years. It has simply expanded as our knowledge of the universe has expanded. It has only been a relatively few years since the whole idea of there being other galaxies was identified. Prior to that, the Milky Way Galaxy was the extent of our collective understanding with the Solar System being the only assembly of planets so attention was naturally restricted to it but that did nothing to hamper the conviction that life must surely exist elsewhere. Mars was a prime suspect and there was great excitement when Giovanni Schiaparelli identified canals on Mars in the late eighteen hundreds. (Scienceblogs.com) This idea was reinforced when Percival Lovel – a well respected astronomer – also thought that he saw the canals of Mars and so he drew a diagram of them. Those were the days when astronomers actually looked through the telescope. This practice has long been abandoned and now cameras are located at the focus points and pictures are taken instead. Large telescopes are very precise instruments and it was always a cause for concern when an astronomer actually climbed onto the instrument and took his/her position. This was commonly at the primary focus at the upper end and the uncertainty of the weight of the astronomer and his/her instruments always necessitated a rebalancing of the instrument. Neither was it a pleasant job. All night long they would be sitting in a very cramped position with their backs exposed to the chilly night air so it was actually a relief when the practice was abandoned. Further the variation in the size and weight of the astronomer was occasionally compounded by the uncertainty of his personal equipment. There was a case involving the giant 200 inch telescope on Mount Palomar in California when an astronomer brought with him an instrument that was too large to fit into the small space at the upper end. At the insistence of the astronomer the instrument was lashed to the outside of the telescope frame and it projected beyond the usual outer limit. In this incident the instrument projected from the telescope for almost one meter (about 3 feet). Things went all right until the telescope was swung around in the darkness of the observatory and the lashed-on instrument crashed into the spiral staircase on the inside of the dome. The instrument came loose and fell to the floor below. It missed the mirror which at that time was the largest one in the world and was practically irreplaceable. The technician was soundly lectured and the astronomer was no longer welcome.

Even as it is commonly the case at the present time that the motivator for any project involving the solar system or beyond is the hope of finding evidence of life so it was the motivator for numerous astronomers at that time (1960's and 1970's). While hope springs eternal there is, from a scientific viewpoint, no hope whatsoever that life will ever be found apart from the Earth.

The basic conviction of numerous scientists at the present time is that the Earth is about 4.5 billion years old. This idea fits in quite well with the accompanying conviction that the universe is some 12 billion years old. This later idea is based in part on the observations that certain objects in the cosmos are so very far away that it must have taken 12 billion years for the light to get from there to the solar system. There is certainly rationale for such ideas but they do depend on several other factors being supportive of the basic idea. One of these factors is the speed of light. The speed of light has been measured repeatedly and forms the basis for many of the notions in astronomy because it is assumed to be a constant. In fact it is commonly referred to as the 'constant' of the speed of light. Constants are important. We do need references for our personal stability so the more constants that can be identified the more comfortable we feel.

Sea level is a constant. Even though actual sea level at any particular location will drift up and down due to tides, winds and currents there is a particular elevation of the surface of the sea that is recognized as 'sea level'. The idea that it is actually changing at the present time due to global warming is upsetting enough for those who will be directly and seriously affected but it is upsetting for everyone (which includes most of us) who needs such things to be perfectly reliable and just stay the same continually. The 'speed of light' is certainly in this category and the thought that is might have been different in the past or might be different in the future is quite unnerving. We need all of our constants to stay constant. If the speed of light is actually changing, numerous other ideas will immediately be seen to be in jeopardy and this would also be upsetting. The 'speed of light' idea is intimately connected to our concept of time. Unfortunately, time is not very well understood. The last person to do serious exploration in this area was Albert Einstein about one hundred years ago. He understood that time would not always be the same for all locations and all circumstances. It is not really a constant of the universe even though the general thinking up to that point was that it was a constant. (Stephen Hawking also wrestled with the time question and has presented a discussion in his book 'A Brief History of Time'.) If this lack of constancy is actually the case, much of what we think about the universe comes unglued. Did it really take 12 billion years for light to travel from remote regions of the universe to our solar system or did it just take two weeks? What would that do to our concept of the 'Speed of light'? To say the least it would be upsetting. Similarly if our Earth isn't really 4.5 billion years old then how old is it? Many derivative ideas are closely linked to this basic notion and would have to be totally reworked if the Earth was much younger. Unfortunately this turns out to be the case which immediately means that a lot of rethinking will be required to once again have a reliable framework for our ideas. Prior to Einstein a clock was a clock and time moved at the same speed no matter where you were or under what circumstances you might be operating. After Einstein that basic way of thinking was abandoned. Something similar will be required when it becomes clear that the Earth isn't really 4.5 billion years old at all.

The question of extraterrestrial life must be approached from a probability perspective. This means that a final, complete and irrefutable conclusion can never be drawn from a scientific viewpoint. This might be disconcerting to some but when a probability is objectively calculated and it turns out to be so infinitesimal that a small leap of faith from there to 'there isn't any life elsewhere in the universe' becomes quite reasonable. Someone can always say 'but'. A probability conclusion might be satisfying to an abstract mathematician or a research scientist but to most of us, including the extremely small step required to draw a firm conclusion will be much more satisfying. In a case like this, absolute certainty could never really be reached unless every star in the universe was investigated. Even then we might get fooled because we could never be certain that something didn't happen just after our backs were turned. Technically therefore, the matter can never be settled. However in order for most of us to live our lives from day to day we must work with more definitive conclusions. While leaving the realm of science fiction behind the present discussion will work from the available evidence and then come to a conclusion based on that evidence.

The common understanding of this age is that the Universe must be teeming with life. This conviction is held despite the fact that there is no evidence whatsoever to support it. The general thinking seems to be that since the universe is so vast with virtually an uncountable number of stars there must also be an uncountable number of planets. There is a certain amount of validity to this way of thinking because of the observation that the galaxy includes a very large number of stars and that there is a similarly large number of galaxies. Simple mathematics therefore indicates that there must be thousands of billions of stars. Reasoning that even if a very small percentage of them had a habitable planet there would be hundreds of billions of possibilities. How could anyone argue with such logic?

There are two related basic beliefs that support and reinforce the above conviction. The first of these is that The Milky Way Galaxy (i.e. our own galaxy) has been in existence for a very long time - that is for several billions of years. In particular it is widely believed that the Earth has been in existence for a period of time between four and five billion years. With so much time available life would surely have had time to develop. This conviction, in turn, is directly connected to the third widely accepted belief, namely that life evolved from very simple forms to more complex forms with man clearly being the 'highest' form (on the Earth). It would not be an exaggeration to recognize that the vast majority of humanity accepts this belief as 'a proven scientific fact' leaving non-believers as being completely out of touch. Far-away planets might even have already evolved 'higher' forms of life (which might be trying to contact us). Such thinking is excellent fodder for science fiction of which a vast supply is available.

These three inter-related and mutually-reinforcing ideas including; (1. The universe includes billions of life-supporting stars; 2. The universe is very old; and 3. Life evolved from lower to higher forms culminating in human beings) are recognized as 'science' making any counter arguments seem foolish. Who can argue with 'science'?

It is therefore going to come as a shocking and surprising development that the above three convictions are not quite as certain as we have been led to believe. Actual science - that is science based on observation and measurement - totally upsets all of these beliefs to the degree that none of them have any credibility whatsoever. Some of these factors have been discussed in 'The non-Myths of the Bible' and in 'The Window of Life'. The comments to follow will relate directly to the question of life on far-away planets. The conclusion that there could not possibly be life at any other location in the universe might be upsetting but sooner or later the Truth must be faced or it will bite us all in the a-s.

2.0   Introduction

To the casual and unaware observer there seems to be an unlimited number of possible locations in the universe for life to exist. However this is not the case. In order for life to exist the complexity of the environmental circumstances that is required to be in place is almost too great to describe. This factor follows the primary requirement which is to find a place where the necessary environmental circumstances could be set up. By place of course we are referring to a planet but even if one were found there might only be a restricted region on that planet where these circumstances might be found. For example, the Earth is doing very well as a life-supporting planet but even in this case the entire planet cannot be used. Antarctica, for example, is cold and ice covered. Nobody can live on Antarctica (without a complete life-support system coming from elsewhere). Not even animals can live there. One must recognize that Penguins go ashore on Antarctica for a significant period of time every year but without the ocean to supply their food and water requirements, penguins could not live there either. There is no food there. There is no water there in liquid form. Plants do not grow there. There are no trees. Humans cannot live there - partly because of the lack of a food supply but also because of the extreme cold. In order for humans to survive on Antarctica all of their life requirements must be imported. Without the necessary provision of food and shelter from else-where a human being cannot live on Antarctica.

Mars is one of the favourite places where it is commonly suggested that human beings could live. Occasionally there is talk of setting up a colony on Mars. Such suggestions fall far short of reality. The average temperature and the temperature range on the surface of Mars is actually very similar to Antarctica. It is correct that there is coal on Antarctica but mining coal to support a colony is not feasible. If it was decided to recover Antarctica coal a complete mining operation would need to be imported. On the other hand the high Arctic islands have a supply of wood. In fact there are massive piles of frozen trees several hundred feet high(187) on some of the islands north of Russia. However there doesn't seem to be any interest in harvesting these trees for a fuel supply partly because the energy required for harvesting would be greater than the energy that could be recovered by burning. The Mars situation is similar in this regard. The energy required to ferry a supply of heating fuel to Mars would be far greater than the heat value of the fuel. While it would be technically possible, it is quite clear that the possibility of a long-term self-supporting colony on Mars is extremely bleak. (Solar panels can provide electricity for lighting but are not really suitable for heating.)

As we are recognizing the heating problem on both of these places Mars has another problem which is completely insurmountable. On Mars there is no magnetic field of any significance to provide protection from cosmic radiation. The Earth has a magnetic field but even in this case protection is diminishing because the strength of the field is falling off. Mars is totally exposed to incoming radiation and this will be fatal to all forms of life on Mars the next time there is a solar flare of any magnitude. There is simply no place to hide from incoming radiation on Mars whereas on Earth we are protected by the magnetic field.

It is very clear from these few realities that in order for a planet to be able to support life a host of basic parameters must be in place. As we proceed it will become apparent that finding a planet with even a small portion of necessary factors will be seen as extremely improbable. After all, probability is the way we must approach our discussion because of the extremely high numbers involved.

3.0   Carbon-Based Life

The argument has occasionally been raised that there must be life in the universe – especially if we allow for forms of life besides our own. While this has a certain ring of logic about it, there isn't any science to support the idea. 'Life as we know it' is life based on carbon. The carbon atom has a set of characteristics which enables the complex formations and transactions that occur throughout the biosphere to exist and carry on with their many and complicated transactions. No other atom exists that can do these incredibly complicated things. Life - even at the simplest level - is complex beyond our currently-available ability to describe. In fact the simplest assembly capable of reproduction involves hundreds of different kinds of proteins, each of which is an incredibly-complex structure in itself. There is no such thing as a 'simple' form of life and carbon is necessary for all of this to happen.

For a time, silicon, the most similar atom to carbon, was promoted as a possible basic atom on which complex life-forms could develop. This idea was abandoned when it was realized that silicon could not really carry out the needed transactions to get the job done. Only carbon could do this!

Neither can appeal be made to some unknown atom or to some unknown set of atoms. The Periodic Table includes the only types of atoms that exist and it is understood that the entire universe has only these atoms available. The signatures for many of them have already been detected in distant stars and the idea that the particular set of atoms that is currently known is the only one that exists has never been challenged. Therefore carbon is it. No other life-form based on anything else is possible. The only escape from this restriction is to enter the realm of science fiction. In this realm anything is possible and the only limitation is one's imagination. Our current pursuit is much more serious however, so the realm of science fiction will not be included.

4.0   Finding a Suitable Sun

4.1   The Galactic Habitable Zone

The universe is where we begin. The universe is a very large place and includes an uncountable number of objects. The most obvious of these are the stars. No one will deny that there many stars in the universe. In the Milky Way Galaxy (i.e. our own galaxy) various people have tried to estimate the number of stars. It goes without saying that the stars cannot be counted. Even trying to count to a million would take more time than anyone could justify but beyond that reality the stars in the galaxy are clustered and grouped in such a way that they can barely be distinguished with a high-powered telescope. So estimates will have to do. At the location in the galaxy where the Earth is located, stars are usually spread far enough apart so that counting is possible. However as the telescopes are trained towards the center of the galaxy even trying to count becomes totally hopeless. There are simply too many and from our perspective they appear much too close together. Just to underline the difficulty, at one time it was declared that the galaxy was home to 100 billion stars. This has been rethought - partly because of improved instrumentation - and partly because of improved methods of estimating. It is now commonly suggested that the galaxy has some 300 billion stars in it. From an ordinary persons point of view this is such a high number that it doesn't really have any meaning. However this will be our starting point.

Just as there is a habitable zone for a planet with respect to its immediate star so there is a habitable zone for a star with respect to its galaxy. (A habitable zone is a location where human or animal life could possibly exist.) The Galactic Habitable Zone is understood to occur at the Co-rotation Radius. 'The Sun is in a special location in the galaxy – the Co-rotation Radius. Only here does a star's orbit speed match that of the spiral arms – otherwise the Sun would cross the arms too often and be exposed to supernovae.The intense radiation and gravitation of a spiral arm would cause disruptions in our solar system just as surely as if we were closer to the center of the galaxy. (Therefore) keeping out of the way of the Galaxy's spiral arms is another requirement of the Galactic habitable Zone. (Consequently) most of the stars in the galaxy wouldn't be able to support habitable planets because their rotation is not synchronized with the rotation of the galaxy's spiral arms.' (203)

(Orbit speed refers to the speed at which an object orbits around a center. The stars in the Milky Way Galaxy orbit around the center of the galaxy and the arms also orbit around the center. However they do not orbit at the same speed which means that most of the stars will repeatedly cross through the arms.)

Staying away from the galactic center has an additional advantage. The center of the galaxy is awash in harmful radiation. Solar systems near the center would experience increased exposure to gamma rays, x-rays, and cosmic rays, which would destroy any life trying to evolve on a planet. (203)

Being at the Co-rotation Radius means that there is only a very narrow circular band of stars that are possible candidates for life support. Further, the spiral arms occupy more than 50% of this region further reducing the actual number of stars that exist in the Galactic Habitable Zone. These two restrictions limit the possible number of life-support stars to a very small fraction of the total that exist in the entire galaxy.

Another and totally independent reason to stay clear of the spiral arms is temperature control. As discussed in chapter 6 below, the Habitable Zone for the Earth with respect to the Sun is a very narrow region where it has been declared that the variation in the orbit of the Earth can only be plus or minus 5% from the present location. If we were just a little further from the Sun, the temperature at the surface of the Earth would drop. The Earth would freeze up and crust over with ice. The dust of the spiral arms is expected to produce the same result. Certain commentators credit this type of development to causing an ice age or even multiple ice ages. While this is not a sufficient condition to generate an ice age (because a great deal of heat is also required) it properly identifies that the Earth must avoid entering 'cold' regions of space.

Numerous novas and supernovas occur throughout the inner regions of the galaxy and we could not venture there and live for any period of time even if we could get there in the first place. Fortunately our location is far from the center of the galaxy at the Co-rotation Radius so we can remove that particular hazard from our list. However this restriction, along with the need to be outside of the spiral arms, dramatically reduces the number of the possibly-supportive-of-life stars to a small fraction of the total that exist and it might be optimistic to suggest that there might be only one hundred million possibilities left. (i.e. 0.003%) After starting with several hundred billion this does seem like a major letdown but this is the reality of the situation. So instead of starting our probability discussion with several hundred billion we can properly start with one hundred million instead. The procedure will therefore be to multiply one hundred million by the various other fractions to arrive at a probability conclusion.

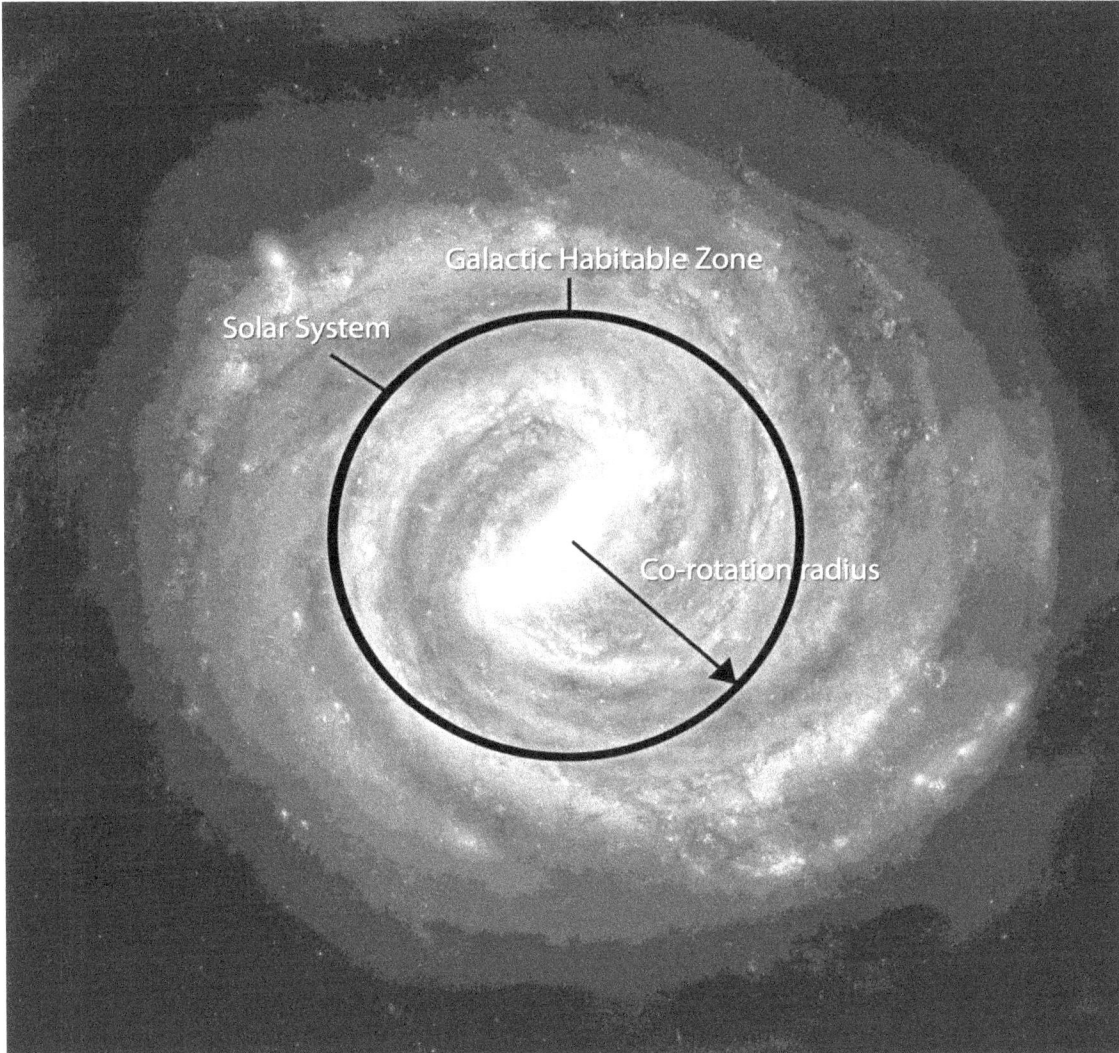

The Galactic Habitable Zone
The Galactic Habitable zone is a narrow region of the Milky Way Galaxy.

The Galactic Habitable Zone is shown here as being very narrow. The further we recede from this narrow zone the greater the difference between the orbital speeds of the stars and the arms and this will determine how often any particular star drifts into one of the arms. Only one incursion would be sufficient to chill any potentially-habitable planet associated with such a star, into a non-inhabitable state. The narrowness of The Zone is therefore justified.

4.2   The Brightness Factor

The brightness factor is one that an be identified quite readily. Stars must have a certain minimum brightness (i.e. surface temperature) to be acceptable as a host sun for a life-bearing planet. The star must be hot. It must have a certain minimum temperature to ensure that any planet which might have even a remote possibility of supporting life can remain at a safe distance. We do need heat from our star but we must receive this heat from a safe distance. The distance must be so far away that the star cannot pull up any significant tide on the planet. If a planet was too close to its star, the gravity of the star would pull up a tide on the planet that would be too disruptive for life to exist on the surface. Large solar-induced tides are destructive enough on their own but they would also cause the planet to stop rotating. This alone would be the kiss of death. If we can imagine a planet rotating and spreading out the heat that it is getting from its star it might be almost within the realm of feasible to consider it as a possible home for some form of life. However, if the planet stopped rotating, one side would then continuously face the star and overheating would result. On the other side over-cooling would result. 'Theoretical models predict that volatile compounds such as water and carbon dioxide, if present, might evaporate in the scorching heat of the sunward side, migrate to the cooler night side and condense to form ice caps. Over time, the entire atmosphere might freeze into ice caps on the night side of the planet.' (121) A narrow strip near the terminator might have an acceptable temperature but with all of the water frozen solid on the cold side the situation would not be worth pursuing any further.

If any planet should form within the thermally-habitable zone of a red dwarf star (i.e. the most common type of star in the universe) it would, within a very short amount of time, become tidally-locked to the star. (204) Then the temperature inequity around the surface of such a planet would render it inappropriate as a place for life to exist. This is the main reason that astronomers reject any star that is below a certain temperature. In order to be in the thermally-habitable zone, a planet associated with such a star would simply have to be in too close where the tidal effect would be devastating for both surface tranquility and long-term heat distribution. In fact, by the time such a planet was discovered, the de-rotation factor would already be in place and the planet would be locked up, with one side continually facing the star. All Red Dwarf stars are in this category so there really isn't any point in considering Red Dwarfs as possible host stars for any life-supporting planet. (The practice continues however mainly because 1, Red Dwarf stars are plentiful, 2, some are relatively close to Earth and 3, it is always exciting to get credit for identifying a new planet.)

The Sun is included in a very small group of stars (approx. 5%) which are bright. (120) This means that the heat produced enables an Earth-like planet to be a considerable distance away and still receive enough heat energy to enable life to exist. At these greater distances, the tidal effect of the Sun is reduced and, in fact, barely noticeable. On Earth for example, the tidal effect of the Sun on the ocean can only be measured in inches.

Unfortunately, most stars are low-mass stars (i.e. Red Dwarfs) and are much dimmer than our Sun. This means that a life-enabling planet must be much closer in order to receive sufficient heat. However closer means that tides will be higher. 'Science Daily reported that "Tides can render the so-called habitable zone around low-mass stars uninhabitable." Astronomers at the Astrophysical Institute Potsdam studied the effects of tides on planets around low-mass stars (the most numerous stars in the galaxy) and found that the ... increased volcanism... make them...uninhabitable. The chance of life existing in the thermally-habitable zone around low-mass stars is virtually impossible due to the tidal effects. In order to find another 'Earth' we first must find another 'sun''. (139) In other words, in order to avoid the tidal problem of having a planet near a dim 'sun', an Earth-like planet must orbit a bright sun. This reduces the number of stars as possible energy suppliers for Earth-like planets to less than 5% of all existing stars. (120)

In spite of such scientific realities, the inappropriateness of dim stars as hosts for life-enabling planets is repeatedly and deliberately over-looked. A recent report (winter 2017) recognized an extremely dim red dwarf star as a possible host for life. In this example it was determined that the host star had seven planets in orbit around it and that at least three of them were in the thermally-habitable zone. In this particular case the point about being too close to the star was deliberately over-looked in favour of the observational advantage. Trappist-1 is an ultra-cool dwarf star which is a good candidate for detecting an Earth-sized planet because when one passes in front of the star, the starlight reaching Earth dips quite dramatically. (175) In other words the observational advantage out-weighs scientific reality. The tidal-reality problem would be particularly severe in a case like this because the planets of interest are understood to be orbiting very close to the star. Trappist-1 is a small dim star but it has six orbiting planets in the thermally-habitable zone. (119) They might sit in the thermally-habitable zone but they certainly do not sit in the structurally-habitable zone! In this case they are so close to the star that they are almost touching it, meaning that the tidal problem would be over-whelming!

It is well understood that Red Dwarf stars are very numerous and hence should provide an abundance of planetary discoveries. Dwarf stars are the most common type of star in the galaxy. This means that there should be many opportunities to find habitable worlds in the Milky Way galaxy. (118) Searching for extra-solar planets is slow, tedious, boring work. However if there is a significant likelihood of finding something, the boredom would be relieved. However, from a scientific perspective this is not valid and relevant scientific realities should not be over-looked in favour of observational advantages.

By eliminating stars that are too dim (I.e. red dwarfs) only about one star in twenty remain as possible candidates. (120) As a probability factor this is 0.05 so we multiply the possible number of stars in the galactic habitable zone by 0.05 and this yields 100,000,000 x 0.05 or 5,000,000.

Certain commentators have suggested that there might be a billion sun-like stars. No restriction of location was placed on this number but if it is referring to the number in the galaxy, which is thought to have about 300 billion stars, it would only represent about 0.000033%. Then this amount must be multiplied by the percentage in the galactic habitable zone, which is probably not more than 0.003% of the number in the galaxy. This procedure reduces the possibilities to about 30,000. While the actual number can never be known, it is clear from these lines of reasoning that the number of possibilities will not be very large. (i.e. 30,000 to 5,000,000)

4.3   The Singularity Factor

A host star must be singular. Multiples of any kind including doubles, triples, and quadruples are to be avoided. The reasoning for this conclusion is that a multiple-star assembly would prohibit any planet in the vicinity from having a circular orbit. A circular orbit is a minimum necessity for temperature stability and it is imperative that if any planet is to be a home for living creatures it must have very tight temperature stability. Even variations of a few degrees would not be acceptable. In fact, the average temperature of the Earth must be exactly where it is at the present time or devastation will result. The current average surface temperature of the Earth is +15C and the concern among many scientists is that it might rise to +17C. Great devastation would follow a rise to +20C. These really are very small temperature deviations which clarifies that non-circular orbits for a life-supporting planet are simply not an option. Consequently star assemblies of two or more stars are not possible candidates for a life-supporting planet. In spite of this, as with the brightness factor discussed above, multiple-star assemblies are occasionally identified as possible hosts for a life-supporting planet. Recently astronomers announced that they had discovered a planet with a mass similar to Earth's orbiting in the habitable zone of Proxima Centura, one of a cluster of three stars in the assembly named Alpha Centura which is the closest star to the Solar System. (117) The other two stars in the cluster are called Alpha Centura A & B and orbit each other about as far apart as the Sun is from Saturn. Proxima Centura is a small red dwarf and the mentioned planet is orbiting Proxima Centura in about eleven days. This means that it is very close to Proxima Centura and that the gravity of the star has locked the planet so that it always shows one face to the star. The hot side will be too hot and the cold side will be too cold. At the mid area any atmosphere will be depleted of CO_2 and water vapor because it will all be frozen on the cold side. There could never be life in such a place. (Since the planet is very close to Proxima Centura the influence of the other two stars on the planet's orbit will be minimal.)

The fraction of stars that this leaves is not exactly known but certain commentators have weighed in on the matter and suggest that most stars are multiples. The frequency of doubles and triples and other multiples is very high. (i.e. among sun-like stars) Astronomers observed all 123 of the Sun-like stars visible to the unaided eye in the northern hemisphere. More than one-half of them were multiples among those that could be observed at the time. (116) These comments were made more than thirty years ago and since then the percentage of multiple stars observed has increased confirming that single stars are very rare and that our Sun is part of a minority population. (115) Other commentators offer similar or higher estimates such as - probably 75 percent of all stars are members of a binary system. (114) In addition to the binaries there are triples and quadruples as mentioned. When these are also included, the proportion of multiple stars is upwards of 80 percent.

Some estimates have placed the portion of multiples at ninety percent. (111) Examples of binary stars within the near portion of the Milky Way Galaxy are; Eta Cassiopedia, Procyon, Sirius, (brightest star in the sky) Ross 614, G1 166, G1 702, G1 783 and Groomsbridge 34. Examples of triples are; Alpha Centura A and B along with Proxima Centura, (the nearest star system to the solar system) G1 166, G1 1245, G1 570, G1 663 A and B along with G1 664 and EZ Aquarii. (110)

It is an absolute necessity that a star be singular before there is any possibility of an associated Earth-like planet having a circular orbit. (Massive spheres of gas such as Jupiter do not qualify in this case as Earth-like). Most stars are binaries. Instead of just one glowing mass of hot gas, there are two. The two stars revolve around each other and from a great distance it appears that there is only one object. It requires the optical advantage of a telescope with its associated instrumentation to determine that a star, which appears like a single object, is really two objects circling around each other. In some cases, there are three objects circling around each other. It is reasonably understandable that two objects could circle each other for a period of time, but it is not really understandable how three objects could do this for any period of time without their orbits becoming unstable. However, whether this happens or not is quite secondary to the fact that two or more massive objects, (like our Sun) which circle each other, have associated with them, both a gravity field and a thermal energy field, which includes massive fluctuations. These fluctuations guarantee that no nearby object, like a life-enabling planet, could ever have a stable, repeatable, nearly-circular orbit and an appropriately-warmed surface. In fact it might not have any orbit at all unless it was close enough to one of the stars so that star's gravity clearly dominated the planet's orbit. It is more likely that ' … untold numbers of worlds have … fallen into their suns or been flung out of their systems to become "floaters" that wander in eternal darkness.' (113) (We note that several 'floaters' have been found recently.)

All of the stars in the vicinity of our Sun are multiples. In particular, the closest star to the Sun is a triple. There is no point whatsoever in considering a planet orbiting in this type of situation as a possible candidate for life support. As a result of all of this observation and scientific comment the probability factor seems to be in the range of 0.1 to 0.2. If we use 0.15 the number of possible life-supporting candidate stars in the galaxy is reduced from about 5,000,000 (i.e. the more conservative above estimate) to about 750,000.

4.4   The Dead Star Factor

Occasionally astronomers refer to a star as having a dead partner. It is never exactly clear what this means. There was a time when heavenly bodies were readily placed into categories. There were stars, planets and moons. Identifying which category to use was never a problem because this is the way that the solar system is arranged. The situation is not quite so identifiable for other assemblies of stars and objects that orbit around them. 'Many are so strange as to confirm the biologist J. B. S. Haldane's famous remark that "The universe is not only queerer than we suppose, but queerer than we can suppose."' (109) A dead star would seem to be an object that was once glowing as a star and then it cooled off and stopped glowing – at least in the visible spectrum. It would not seem appropriate to simply call it a planet because planets are not hot – not that hot anyway. A probability will not be suggested for this factor but if any dead stars exist near an otherwise appropriate star it only further reduces the number of possibilities of finding a suitable host star.

In any event if we review the factors from the above categories we have; Probability = 100,000,000 x 0.05 x 0.15 x 2 other factors = <750,000. In other words, from the above factors, the probability is that less than 750,000 stars in the Milky Way Galaxy might be appropriate to support a life-bearing planet.

4.5   The Sun Size Factor

A host star must be in a certain size range. 'Too small' means that the star is too young (i.e. according to nuclear theory). 'Too large' means that the star is too old. Stars are understood to increase in size as time goes by so an old one would probably not last long enough to allow life to develop or if it had developed it would probably have already died out. (i.e. Some commentators suggest that intelligent life would annihilate itself within a few centuries of obtaining the atomic bomb. (108)) 'Too small' would suggest that there hasn't been enough time for life to develop. For example, our Earth is thought by some to be 4.5 billion years old and that human life has only just arrived. Therefore any star much smaller (i.e. younger according to the theory) would probably not be a good place to investigate because any small life forms would not yet be ready to show themselves. All of this type of reasoning derives from the thinking accompanying the Theory of Evolution which must be accepted if one is thinking in terms of billions of years. Of course if one does not think in those terms, shorter 'temperature windows' might be acceptable.

The complicating factor accompanying such extremely long-term thinking relates to the theories of how a star operates. Nuclear fusion is the prime suspect and with nuclear fusion a star would be expected to steadily increase in both size and brightness (i.e. heat output) as time goes by. Our Sun, for example, is thought to have been about 25% cooler 4.5 billion years ago which begs the question of it being a good source of heat at that time because it would not have been hot enough to have kept the Earth from being frozen solid. If the Earth was ever frozen and covered by a layer of ice, the albedo or reflectivity factor would have been very high. Most of the incoming heat from the Sun would have been reflected away and very little would remain as a source of heat for life. How then would it ever thaw out? There isn't even enough heat from the Sun at the present time to keep the surface temperature of the Earth above freezing if it wasn't for the greenhouse effect. We need both the current input of heat from the Sun as well as the greenhouse gas inventory (to retain some of this heat) just to keep the surface temperature of the Earth where it is at the present time with very little deviation from this point being allowable. In fact without the greenhouse factor - which would not be fully in place until the Earth actually thawed out and released some water vapour into the air - the temperature of the Earth would still not be above the freezing point and would not be for another two or three billion years. This complication really does present a conundrum for 'long-agers' from which it is difficult to exit.

In any event certain commentators suggest that the size of any star that is to be a possible host for a life-supporting planet must be 'in the range between 0.83 and 1.2 solar masses in order for an orbiting planet to avoid either a runaway greenhouse effect or a permanent ice age, as Mars would experience if it had more water.' (177)Hot singular stars come in a complete range of sizes. This means that even though a star might be appropriately hot, it must also be in a very particular size range. This reduces the number of possible stars to a fraction of the number remaining from the previous restrictions. From this factor it is unlikely that even 1% of the remaining stars would qualify, placing the number of possibilities at 75,000.

4.6   The Solar Stability Factor

Stars are not necessarily stable. Sometimes they explode as novas. Ideas have been advanced to try to explain these explosions and some of them apply to multiple star systems which this discussion has already covered. Also as stars get old they are thought to become unstable but 'old' is a relative term which is difficult to define with any precision. Further, it is doubtful if all types of unstable solar behaviour is really understood. Some probability might be attached to this factor but it will not be attempted here. In any case it would reduce the number of possible candidate stars even further.

4.7   Stable Solar Mass/Size/Heat Output

In order to be a suitable candidate for a life-supporting planet, a star must have a stable mass over the long term. Unfortunately, solar masses are not necessarily stable. Stars have an enormous amount of mass. That is, they consist of a great amount of material. The amount of material in any given solar mass determines the orbit of every object that has been captured by it and every object that continues to endlessly orbit around it. In fact, it is the orbital features of an orbiting object that are used to determine the mass of the host star. Therefore, if a star has a planet in orbit around it, the length of that orbit and the time it takes to make a single revolution can be used to determine the star's mass. The same technique can be used to determine the mass of a planet. If a planet has a moon, the characteristics of the moon's orbit can be used to determine the mass of the host planet. Determining mass in the absence of an orbiting object can also be done but it is more difficult. However, if a star has an orbiting planet, the mass of the host star can be immediately calculated.

A problem arises if the host star does not have a stable mass. If its mass should increase, for example, the orbit of a planet would change and in fact it would spiral inward into a smaller orbit. This would not be satisfactory for any planet that was the host for either animal or plant life of any kind. While achieving the proper orbit is most improbable in the first place because there is such a small margin for deviation, it is a wonder that even the Earth can stay in the very narrow habitable zone (of the Sun) that is so essential for life.

The orbit of the Earth is referred to as being a Goldilock's orbit. It is 'just right'. (181) It is slightly elliptical and this works synergistically with the surface material of the Earth to provide the best possible conditions for temperature stability in exactly the right range. Orbital deviation is not permissible. In fact it has been determined that even a 5% deviation would spell disaster and the surface temperature of the Earth would be outside of the permissible range for life to exist here at all. (178) What would happen then if the mass of the Sun was increasing and causing the Earth to spiral inward closer to it or what if it was decreasing and allowing the Earth to spiral out farther? Unfortunately, observations for the last 30 years or so indicate that the Sun is becoming smaller and dimmer. (179) This is happening in spite of the observations that indicate that more mass is falling into the Sun on a daily basis. If the reduction in size indicates a reduction in mass, the Earth **will** drift outwards and become steadily further away. Dimmer indicates that it is losing heat. The combination of a reduction in heat output and a reduction in mass means that the Earth **will** drift outwards and will soon be located outside of the thermally habitable zone (which would, at the same time, be moving closer to the star). In other words, the Earth will become too cold to support life and in fact all of the moisture and soil at the surface will freeze solid. Once a layer of ice and snow appears on the surface, the temperature at the surface will lock up well below freezing and recovery will be impossible. (180)

The gravity of the Sun is great and it continually attracts more and more material into itself. One staggering example is the comet that went into the Sun in 1979. This comet was a monster and various reports suggest that its size was larger than the Earth. (101) That might not represent very much mass in comparison to the enormous mass of the Sun but what would be the result if this type of activity continued to happen over one billion years? However, it does continue to happen with comets falling into the Sun on a regular basis. Other material also falls into the Sun and it is a wonder that even more isn't pulled in as well. The gravity of the Sun is enormous and very difficult to resist. This makes it hard to send satellites out of the solar system because they must have enough energy (i.e. speed) to overcome the pull of the Sun's gravity. It would be even worse for any spaceship returning from an interstellar journey because it would have to resist the Sun's gravity all of the way in through the solar system or it too would be pulled in. If a returning spaceship were to enter the solar system at some small fraction of the speed of light it would have to apply the brakes full time just to avoid going right into the Sun at full speed.

The Sun also loses mass. Every time that a solar flare bursts out, some mass is ejected outwards. Over the course of a year none of these activities would matter but over very long time frames they would matter a lot. The Sun must retain a constant mass in order to keep the Earth in a stable nearly-circular Goldilock's orbit. The Earth cannot be allowed to spiral inwards (due to an increase in solar mass) by even a small amount or it would overheat. Similarly it cannot be allowed to spiral outward or it would become frozen solid and remain frozen solid. If it has been gradually spiralling inwards for 4.5 billion years it must have been further away in that distant past. The Sun (according to nuclear theory) would have been dimmer at that time and the double factor of being further away and having a dimmer Sun would have ensured total long-term freeze-up. Since this has not happened the declaration that the Earth has been in existence for a long time is cast into doubt. On the other hand, if the Sun is slowly losing mass (as apparently has been observed) the Earth will just as slowly drift further away.

 Getting the Sun set up as a stable long-term source of heat for the Earth seems like a very difficult criteria to meet. The same criteria must apply for all other possible candidate life-supporting stars in the universe. A probability factor will not be applied to this situation but it is clear that the difficulty in getting long-term stable mass for a star (to be an appropriate host for a life-bearing planet) will only reduce the number of possibilities. The previous criteria reduced the number of possibilities to about 75,000 so adding the requirement for long-term mass stability will only reduce that number further.

4.8   Review

The first three factors mentioned (i.e. The Galactic Habitable Zone, The Brightness Factor and the Singularity Factor) enable credible probability factors to be applied. When this is done it is readily seen that (while there certainly are a great many stars in the galaxy) there are only a relatively few stars that meet these three factors alone. In fact, even with full recognition that there are probably 300,000,000.000 (or even more) stars in the Milky Way Galaxy, there are less than 1,000,000 that meet these first three basic factors. Things do not improve from here.

The Sun-Size Factor reduces the possible number even more. Suggesting that only 1% of the above number would qualify would be realistic. Recognizing the Dead Star Factor and the Solar Stability Factor imposes further restrictions. Unfortunately, this leaves us in the range of less than 10,000 stars. Another important factor must also be recognized. While the size of a star must be in a certain size range, that size must be stable. It must remain stable over the entire period of interest. Unfortunately, we cannot expect that stars will remain at any particular mass. They increase in mass due to their incredibly strong force of gravity. Material from the neighbourhood of the star is continually being attracted into the star. This includes both rocky asteroid-like material as well as comet material. Material also leaves. Every solar flare ejects material away and the enormous heat output suggests that fuel is being consumed. Will these factors offset? Any response would be speculative so a definitive factor will not be applied but it is readily seen that a serious uncertainty has been recognized. Even when we restrict our discussion to the more definitive factors the following conclusion is readily reached;

There are very few stars in the universe that might be possible candidates to provide the appropriate level of both thermal and gravitational stability that a planet would require for human and animal life.

5.0   The Void Solar Proximity Zone

The next factor to be considered is the region immediately around the candidate star. In this case immediately will refer to the region out to the habitable zone and beyond for a distance of about another four or five times the distance from the star to the habitable zone. For example, if the thermally habitable zone was 100 million miles from the host star, the distance out to about 500 million miles must be virtually void of planets of any significant size. The reasoning for this is that large planets within this zone would have a detrimental effect on the circularity of a potentially-habitable planet's orbit. An orbit that is close to circular enables the heat received from the star to be constant. A potentially-habitable planet absolutely must have constant heat input. The surface temperature of any planet that is trying to support life must be about one-half way between the freezing point of water and the body temperature of animals. (The body temperature of animals is very close to the upper temperature at which seeds can germinate.) Even seemingly small deviations from this are seen by many scientists as detrimental. The gravitational influence of even one large nearby planet on a small planet like the Earth would not just cause its orbit to be non-circular but could even fling a small planet right out of the system. 'Even if an Earth-type planet existed in this system (i.e. a bright star called Upsilon Andromedae in the constellation Andromeda c/w three orbiting gas giants) it might risk being obliterated ... ' (113)

Recently, a report was filed that 'Earth 2.0 has been found. ... Its sun is about half the size of our own, and it completes its orbit in 130 days. It has four neighbouring planets that are closer to the star but Kepler 186f is the only one located in the Goldilock's zone, with conditions allowing for liquid water – and therefore life – on the surface,' (195) The size of the other four planets was not given but concern is immediately raised that they might be large and therefore prevent the planet of interest from having a stable circular orbit. Also the size of the star indicates that it is a red dwarf and hence not an appropriate candidate for a life-supporting planet anyway (because of both tidal locking and the impossibility of having a moon).

Jupiter is approximately 483 million miles from the Sun but Jupiter with its enormous mass is understood to be a stabilizing factor for the orbit of the Earth. 'Indeed, we may owe our own existence to Jupiter, whose massive influence has, over billions of years, served to stabilize the orbits of our planetary neighbours, making the solar system safe for life as we know it. If

Jupiter's orbit was as eccentric as some extra-solar planets, Earth might never have had a chance to evolve a stable climate.' (165) At the same time it is not really a stabilizing factor for objects in its vicinity such as asteroids, as it is credited with modifying their orbits and causing them to move further into the region near the Earth. (152) It would also disturb the orbits of any other object that drifted too close. The best arrangement for monsters such as Jupiter is to keep them at a safe distance to avoid trouble altogether.

While the presence of a monster planet like Jupiter (at an appropriate distance) must be recognized as necessary to enable a candidate planet to have a stable orbit and therefore possibly be habitable, a probability factor will not be applied.

6.0   Finding a Suitable Planet

6.1   The Habitable Zone

Whenever a far-away planet is discovered, the first question that is usually raised is; Is this planet in the habitable zone of its star? There simply isn't any reason to proceed further with any investigation if a newly-discovered planet is outside of the habitable zone. (In this case we are thinking in terms of the thermally-habitable zone of the host sun. There is also a structurally-habitable zone.) A planet absolutely must be in a location where it receives an adequate amount of heat but not too much. 'If they (i.e. astronomers) manage to discover a rocky planet roughly the size of the Earth orbiting in the habitable zone – not too close to the star so that the planet's water has been baked away, nor so far out that it has frozen into ice - they will have found what biologists believe to be a promising abode for life.' (154) In the case of the Earth even though it has numerous mechanisms to regulate and distribute heat, its habitable zone with respect to the Sun is still very narrow. Just as there is a galactic habitable zone (which is at a very particular distance from the center of the galaxy (153)), so too there is a very particular location for a planet to be in order to receive adequate, but not too much, heat from its host star. (i.e. the habitable zone)

The habitable zone of the solar system is exactly where the Earth is located. This zone is very narrow and it is understood that if we were only 5% closer to the Sun, the Earth would overheat and become uninhabitable. (104) Overheating is easily achieved simply because the acceptable average surface temperature for the Earth is about half way between the freezing point of water and the body temperature of animals and it cannot be allowed to deviate from this level by more than a few degrees. (In particular, the temperature could never be allowed to rise and remain above the body temperature of either animals or people. Also, at such temperatures seeds would not germinate.) While the temperature of the Earth is currently at the right level (i.e. 59F or 15C (159)) there is serious concern that it is rising due to too much greenhouse gas in the atmosphere. The rise anticipated over the next one hundred years is only a few degrees but if it became more than this, wide-spread disaster would be expected to follow. Even a cursory survey around the near universe will confirm that a temperature change of that magnitude is really quite small. This is of very little comfort however if disaster really does accompany such a small increase. While the habitable zone is very narrow to begin with, it is well understood that all of our other temperature regulation factors must also be fully functioning because simply being in the habitable zone might not be sufficient to keep the Earth habitable. While being 5% closer to the Sun is considered disaster being a similar distance further away would also be disaster as the Earth would be much too cold. This requirement is partially recognized in the following comment. 'In 1978, an astrophysicist named Michael Hart made some calculations and concluded that Earth would have been uninhabitable had it been just 1% farther from or 5% closer to the Sun. ... The figures have since been refined and made a little more generous, 5% nearer and 15% farther ...' (104) While the commentator in this case clearly recognized the narrowness of the habitable zone in general, being 15% further away would result in the Earth receiving about 25% less heat. This would not be acceptable because the Earth would simply freeze up. (100) Other commentators have also recognized the narrowness of the thermally habitable zone. 'Recent studies show that even Earth just barely qualifies as a suitable abode for life. If the planet Earth had been placed in an orbit only 5% closer to the Sun, a runaway greenhouse effect could by now have turned the planet into a hothouse – with surface temperatures near 900F, the condition that now exists on the planet Venus. (Barrow and Tipler 1986). On the other hand, if the Earth was only about 1% further away, runaway glaciations would by now have enveloped the Earth with ice ....'(162) While glacier formation requires heat and therefore would not happen if the Earth chilled, the idea that it would become ice-covered and hence uninhabitable, is well recognized by such comments. (We recall that being covered with ice is not an Ice Age but rather a 'snowball Earth'.. See 'The Window of Life' and 'The Asteroid Theory of the Flood and the Ice Age'.)

Drifting out of the habitable temperature range could happen more readily than one might at first think because of the various viscous cycles that would cut in to exaggerate any initial change. For example if the Earth should suffer a chill on a world-wide basis (for any conceivable reason) some significant fraction of our current inventory of water vapor would be lost from the atmosphere. Since water vapor is our most influential greenhouse gas, the temperature would fall even further due to this loss. (160) In fact it would spiral down until any further loss (of water vapor) would not result in any further drop in temperature. Unfortunately, the final temperature would be about -25C. (160) This is the main reason that any outward deviation of the Earth's orbit must be limited to only a very few percentage points.

The great danger to the Earth resulting from chill relates to the formation of an ice-covered surface. The light and heat reflectivity resulting from ice and snow formation is dramatic. Ice and snow 'reflects over 80% of the incident sunlight. The albedo is around 0.8 or 0.9.' (167). (i.e. 80 or 90% of incoming heat and light is reflected.) By contrast 'The albedo of the ocean is less than 0.1, its more like 0.07.' (167) Similarly the albedo of forest is about 0.2. (168) This means that these surfaces are absorbing between 80 and 97 percent of the heat and light that is striking them. If chilling occurred over any significant percentage of a planet's surface, that planet would never be able to warm up (and release some water vapour to further assist with the warming.) and would form a 'snowball' planet as mentioned elsewhere. These realities readily illustrate that in order for any planet to be habitable, no significant fraction of the surface can be allowed to drop below freezing and become snow or ice covered even if that planet was within the habitable zone of its host star.

A planet's distance from its host star is the most obvious and easiest factor to identify but circularity of orbit is closely connected. The orbit must be very close to circular to ensure that the strict distance limits are not violated. While the orbit of the Earth is very slightly elliptical, (i.e. + and - less than 2% from circular which turns out to be 'a stroke of very good fortune'. (172)) its very slight non-circularity works together with the Earth's axial tilt as well as both the distribution and type of surface material to provide an optimum heat control arrangement. For example, when the Earth is closer to the Sun the southern hemisphere is having summer. This is most convenient because when it is summer in the southern hemisphere the large ocean surfaces are facing the Sun more directly. The water is therefore able to absorb the extra heat and release it later thereby helping to regulate the temperature in a most advantageous manner. Also, when it is summer in the northern hemisphere, the Earth is slightly further away from the Sun which means that the larger land areas will not tend to overheat. Land is much more readily warmed than water but land cannot store the heat for later use nearly as well as water. If the northern hemisphere was closer to the Sun (during its summer season) instead it could overheat which would reduce the overall habitability of the Earth. It is difficult to imagine a more favourable arrangement than the one that the Earth has and it is one that is certainly improbable.

An elliptical orbit could carry a planet into and out of the habitable zone even if the average orbital radius was well within it. This would ensure that the planet would not be suitable for life support. A supposedly-habitable planet cannot become too cold for part of its year and too hot for another part of the year. This means that a potentially-habitable planet must not only have an orbit within the star's thermally habitable zone, it must also have an orbit which is close to circular. In the case of the Earth the deviation from circularity is less than 2% outward and less than 2% inward. (105) With 5% as the maximum allowable it seems most fortuitous that the Earth not only meets the requirements but seems to have a slight cushion to spare.

Orbital stability is also a must. The orbit of the Earth is slightly elliptical as mentioned and the Earth follows the same orbit every year. When it is summer north of the equator the Earth is slightly further from the Sun. When it is summer south of the equator the Earth is slightly closer to the Sun. Every year for hundreds of years this orbit has been repeated. This implies that it is being stabilized and it is being stabilized! 'This eccentricity ... is small ... and varies (slightly) because of the influence of other planets.' (164) In particular it is understood that Jupiter provides a significant stabilizing factor for other planets. 'Indeed we may owe our own existence to Jupiter whose massive influence has ... served to stabilize the orbits of our planetary neighbours, making the solar system safe for life as we know it. If Jupiter's orbit was as eccentric as that of some extra-solar planets, Earth might not ever have had ... a stable climate.' (165) 'We should consider ourselves lucky that Jupiter ended up in a nearly circular orbit. If it had careened into an oval orbit, Jupiter might have scattered Earth, thwacking it out of the Solar System. Without stable orbits for Earth and Jupiter, life might never have emerged.' (192)

While it seems to be well recognized that we really need Jupiter it seems reasonable to expect that Jupiter's orbit must also be stable. This is where the other giants of the Solar System come into the picture. Saturn, Uranus and Neptune are also giants and they also have very-close-to-circular orbits (as does Jupiter). Kepler, one of the early scientists involved in explaining the Solar System to us, recognized the harmony of the Solar System. In fact he referred to the 'harmony of the spheres' in some of his writings. This is recognition that the entire Solar System must be in place and operating as a system instead of being an assembly parts moving independently. The orbit of the Earth absolutely must be stable, repeatable and very close to circular in order for the Earth to be habitable. Apparently Jupiter is a necessary factor in keeping the Earth's orbit stable but this could only happen if Jupiter's orbit was also stable, repeatable and close to circular. It appears that the other three giants of the solar System

(Saturn, Uranus and Neptune) are involved in Jupiter's stability which is suggestive that the entire setup is operating in a harmonic inter-dependant manner. What are the chances that another Solar System (with a multitude of planets operating in a similar inter-dependant manner) ever be found?

Furthermore, the Earth has the Moon and the combination of Earth-Moon is recognized as a double planet. This arrangement also contributes to orbital stability similar to the way that a rotating football has greater stability and a more predictable trajectory than one that is just tumbling haphazardly. If you want a football to go to a certain very particular location, throw a 'spiral'.

It really does seem that the possibility of a non-circular orbit is very high and it seems like Earth could very easily have had a non-circular orbit in which case this discussion would never have commenced. What then can be hypothesized about some hopefully-habitable far-away planet? Why would we expect it to have a stable circular orbit? Does it have a large moon nearby and a jupiter at just the right distance (with a stable and stabilized orbit) to provide orbital stability? And what will be stabilizing the orbit of this jupiter?

Jupiter is a monster by any comparison but the other planets including Saturn, Neptune and Uranus are also very large. It seems that in order for Jupiter to have a stable circular orbit they too must have stable circular orbits and ultimately for Earth to have a stable orbit. It really seems that the entire cohort of planets is working together to keep all planetary orbits stable - not just the Earth's. An even more subtle implication is that the other planets are necessary and that the entire package of planets is necessary for the orbital stability of Earth to be possible. As mentioned, in order for any planet to be habitable it must have a stable, repeatable nearly-circular orbit within the respective star's thermally-habitable zone. This in turn appears to necessitate that other planets (apparently large ones) be present and contribute to the orbital stabilizing function. This will be a very demanding criterion for a far-away planet to meet!

Occasionally the habitable zone for our solar system is claimed to be so wide that it includes Mars as well as most of the space in towards Venus. (193) However if the Earth were located much closer to Venus it would be receiving much more solar heat than it does now. The Earth's barely-allowable limit for temperature increase (from the 5% orbital variation mentioned above) is only about 10% whereas in closer to Venus' orbit the increase would be closer to 100%. Clearly this would not be acceptable. Earth's current average temperature of about +15C (59F) would rise until it was well above the temperature at which seeds can germinate as well as the body temperature of animals. The currently-inhabitable areas would all be over-heated and even the polar areas it would be too warm. Just to make matters worse, if some areas near the poles were within an acceptable temperature range they still wouldn't be able to support life because there isn't enough light at those latitudes to grow plants. (201) Consequently any possibility of survival for either plant or animal life over the entire world would disappear.

At the other extreme it is well understood that Mars is so cold that the temperature seldom rises above the freezing point of water. 'With its distant orbit – 50% further from the Sun than the Earth – and slim atmospheric blanket, Mars experiences frigid weather conditions. Surface temperatures typically average about -60C (-76F) at the equator and can dip to -123C near the poles. Only the midday Sun at tropical latitudes is warm enough to thaw ice on occasion. But any liquid water formed this way would evaporate almost instantly because of the low atmospheric pressure.' (194) Therefore neither plant life nor animal life would be possible. For these reasons (at least) Mars is not habitable and will never be seriously colonized.
It is increasingly clear that maintaining an appropriate temperature to enable life to thrive on Earth (even with all of its temperature-regulating mechanisms) is going to be a serious challenge over the next millennium. So trying to include either Mars or Venus in the habitable zone of the Sun seems most dubious.

When remote exo-planets are declared to be in their respective star's habitable zone one would naturally think that the astronomers are thinking in terms of habitable for animals and humans. Unfortunately this is not the case. In spite of the fact that all forms of life are based on carbon and would appear and function the same way that animals do on Earth, 'habitable' is extended to include extremophile forms of life. While this is valid to a certain degree one must be careful to recall that what appears as extreme to a human being isn't extreme in the broader sense. The worms that live near the hot vents on the bottom of the ocean are an example of 'extremophile'. One end of these worms is in the cold water of the bottom of the ocean (about 4C) and the other end is in the hot water flowing up from beneath the ocean floor. Such an arrangement would not be the least bit satisfactory for any of the common forms of animal life and therefore to call it extreme is reasonable. However, even in cases like this, the creature must have protection from ultraviolet light as well as the other types of deadly radiation that emits from stars. They must also have a supply of nutrients even though the amount might be very small. In other words the complexity of the supportive environment, even for extremophiles, means that the vast majority of far-away planets must be eliminated from any 'habitable' list because they simply are not in a thermally-habitable zone nor would they have any reasonable supply of nutrients for any type of life.

While it is clear that being in a thermally-habitable zone of a sun is a necessary condition for the survival of life on any planet, there are always other critical factors involved as well. One might add at this point that this is the basic reason that probing for life on any planet orbiting a Red Dwarf (i.e. a small reddish star which includes 90% of all stars) will be an unproductive task. As has been pointed out by various astronomers, a planet orbiting a Red Dwarf Star in its thermally-habitable zone would have one side permanently tidally-locked to the star. Any water or carbon dioxide that might have existed would be locked up on the cold side making the narrow thermally-appropriate zone uninhabitable. We are reminded of the comment made above in The Brightness Factor discussion 'if you want to find a second Earth, it seems you need to look for a second Sun'. (139) The Sun is included with that very small portion of the stars in the universe that are considered hot enough to have thermally-habitable zones that are far enough away to prevent any planets, that are in those zones, from becoming tidally-locked to them.

Further, if the Earth was in the thermally-habitable zone of the Sun, (as it is at present) if it was sitting upright because it did not have an Axial Tilt, (see 6.3.2 below) it might be expected to have a thermally-habitable region between ten and twenty-five or thirty degrees of the equator but the extremely dry air that would be continually descending in this area would render this region desert and hence uninhabitable.

It is very clear from this basic discussion that any planet that provides an environment where temperature is not regulated in a manner similar to the way that it is done on the Earth(i.e. by numerous factors being in play simultaneously) will not be acceptable as a life-supporting location and the search should simply move on.

6.2   Structural Integrity

Even as a house or a bridge or a tall office tower must have structural integrity in order to be useful, so too the structure of any potentially-habitable planet must also have integrety. The surface cannot be fluid, nearly fluid or composed of solid blocks floating and bobbing in moving fluid. Neither can it repeatedly split open. It absolutely must be predictably solid before any form of animal life could possibly become established.

Before moving on we must recognize that while there is a thermally-habitable zone associated with all stars there is also a structurally-habitable zone. This second necessity for habitability is very seldom recognized as a habitable criteria but it is just as important as the thermal factor. The surface of any life-supporting planet must be structurally stable and reliable. How could any buildings be constructed or roads built if the crust of the Earth was in constant turmoil? Even on Earth there are places where roads, buildings or bridges can never be placed. One of these areas is the Amazon River valley in South America. The Amazon is probably the most powerful river on the Earth when both the flow rate and the volume of water are considered. Along the shores in certain areas towns exist but they must be located with full recognition of the variability of the Amazon's flow volume. Both the depth and width of the river can change dramatically within a few hours. On one occasion a report was received that at sundown the river appeared to be less than one mile across. At sunrise the water level had risen twenty-five feet and the width had increased to about two miles. With such dramatic variability and with the general lack of structural stability of the shore material, there will never be a proposal to build a bridge across the Amazon. Another example of an area with structural limitations is a fault line where the land on one side is moving very slowly with respect to the other side. Would anyone ever build any building across a fault line like this?

Just to review, in order for a planet to be structurally habitable, it must be so remote from its host star that the pull of gravity of the star will not continually distort it's surface material. This means that the star must be hot so the planet can stay well back and still receive enough heat. These criteria exclude 95% of stars as possible candidates to support life because they are just not hot enough. To satisfy the structural criteria a planet must be remote enough from its star so that the pull of gravity of the star is reduced to a level which allows for structural stability of the planet's surface. This factor is of particular interest on the Earth because the Earth has a very thin crust which can barely withstand the gravity pull of Sun and the Moon at the present time. Any more tugging would destroy the structural integrity of the crust which would turn into a group of bobbing, grinding plates and nothing would be able to live on the Earth.

In spite of these restrictions red dwarf stars are continuously presented as possible candidates to support life. At the same time it is admitted that these planets, while in the thermally-habitable zone of their host stars, have suffered from the gravitational pull of these same stars to the degree that they would be tidally, permanently distorted. As mentioned above, this means that even if they had rotated at one time, they would have stopped rotating and one side would be continuously locked in the direction of the star. An example of this is given by Gliese 581c. 'It is tidally locked (always faces the parent star with the same face) so if life had a chance to emerge, the best hope of survival would be the 'twilight zone.' (121) Unfortunately this type of hope is no hope at all. When one side of a planet is always facing a star, that side will be much too hot. At the same time the other side will be much too cold. In the twilight zone the surface temperature might be in the habitable range but this does not necessarily make the planet habitable. If there had ever been an atmosphere, any volatiles like water would boil on the hot side and migrate to the cold side. On the cold side they would freeze solid and thereby become locked out of circulation altogether. This would also apply to carbon dioxide which, if any had been present, would become locked up as ice on the cold side as well. (121) Therefore just being warm enough in the twilight zone would not make a planet habitable and red dwarfs should not be considered as possible candidates for a life-supporting planet.

6.3   Temperature Control

Very strict temperature control is paramount for any location to be habitable. While we are primarily interested in planets at the present time, the need for temperature control is easily seen to be absolutely necessary for any situation deemed appropriate for human life. While human beings are widely spread across the Earth there are many locations where they live that cannot really be considered habitable. In all such cases local control must be maintained. Shelter from the weather is necessary - particularly cold weather. It is true that explorers have gone to both the Arctic and the Antarctic and have returned (in most cases) to tell about it. However none of them did it without bringing along shelter. People still venture into both the Arctic and the Antarctic but they all take with them some form of shelter. Tents, igloos and boats are common. Within these shelters heat is required. Water will be frozen so it must be melted before it can be used. Feet will freeze if they are not protected properly. Most types of food are uneatable if frozen. There is no argument that if the temperature is very much below the temperature of a typical living room both heat and shelter must be available before the location could be considered habitable.

Just to emphasize this point closer to home, even a house in a temperate climate must have temperature control. Consider the case of a new, properly-insulated house in a town in the northern USA. This house has high-grade windows. The walls are insulated to the latest standards. The furnace is brand new. However we cannot live in this house without a front door. If the front door is missing, the furnace could work all night and the temperature in this house might not rise at all. In order to achieve temperature control all of the relevant factors must be addressed.

Consider the case of the early settlers and pioneers of this country. There have been many cases of people going into areas where nobody had previously settled. When darkness descends and the temperature drops a fire must be lit. While this will offset the cold for the initial stay, a shelter will become mandatory before long to achieve temperature control of even a modest living space. Log cabins have frequently been built to enable the development of a minimal heated space. The fire in the stove must be kept burning. Whether coal or wood was used, the stove had to be kept fed with fuel. When wood was used, this usually meant adding more fuel during the night.

While these examples point out the importance of temperature control, we are already starting off with a situation that is far from the extremes of the universe. It is possible to survive on the surface of the Earth because the Earth has numerous temperature control factors already in play. The same cannot be said for our neighbors throughout the solar system. Mars is extremely cold. Venus is extremely hot. Mercury is even hotter. There is no point in discussing the outer planets because their temperatures are so extremely cold that it is totally inconceivable how temperature control for any form of animal life could be achieved. If a manned mission should ever venture to any of these other locations in the Solar System, temperature control within the living space will be requirement number one. The Earth is blessed with a series of factors which are working synergistically to keep the Earth on the average, (even though it is situated in a very cold universe) within a livable (for human and animal life) temperature range. One factor could not do this. Numerous factors are required. Our discussion will begin with the basic planet itself.

The temperature at the surface of the Earth is far from being an accidental development but rather involves several factors working together including; atmosphere (Greenhouse Effect), axial tilt, the ocean, the elliptical nature of the Earth's orbit and the structural/thermal nature of the planet, to enable habitability.

6.3.1   The Warm Earth

The Earth is warm inside. Several factors indicate that this is the case. First consider the situation down in a deep mine. There was a nickel mine near the city of Sudbury Ontario. Since nickel is a much sought after metal, the mine was extended deeper and deeper into the Earth until it was over a mile down. The mine workers had to work at that level but could not unless cooling air was brought into the work area. It was simply too warm at that depth and the temperature was well above normal room temperature. (124) Generally speaking the temperature goes steadily up as we go deeper into the Earth.

Other indicators tell us the same thing. A hole was drilled into the Earth in Russia on the Kola Peninsula near Finland (called the Kola Borehole). The drilling crew was very persistent and the project kept going for about twenty years. It would have kept going even longer but could not because the bit was over-heating. The temperature had risen to 180C (126) which was too high for the steel in the bit to retain its structural integrity. It was about to soften. Rock at such a depth would have been just a little below glowing dull red and the hardness (& viscosity) of the rock would be about to start dropping. Within another few hundred degrees the rock would be quite pliable and not be able to hold any particular shape.

A situation like this could be compared to drilling a hole in the ice on Antarctica. The research on Antarctica includes drilling deep into the ice. While a hole can be opened in the ice by this method the hole will not stay open. (127) The ice, of course, is under tremendous pressure and since ice is not really very strong, it will start to shift and soon the hole will become distorted and as more time goes by the hole could close altogether. Actually the drill could be lost if it wasn't withdrawn right away.

Another example which indicates that the interior of the Earth is hot is volcanoes. Volcanoes pour out red-hot molten rock. Within a few days this molten rock will become solid and material that could not support a small boulder when it came out of the interior of the Earth will be able to carry heavy loads. When it came out of the ground it ran like water. When it cooled down it could carry heavy machinery such as a bulldozer.

Some have argued that while the material directly beneath the crust is certainly very warm and molten, further down it becomes solid again supposedly due to the tremendous pressure of the over-burden. This is immaterial. The temperature of interest is the temperature right under the crust. The Earth is a very large sphere of material and comparatively speaking the crust is very thin. If the material immediately under the crust is hot it is appropriate to recognize from this that the surface of the Earth is enjoying a warm interior. From this warm interior heat steadily rises to the surface and escapes into space. However the benefit remains. The Earth has a warm interior and this provides an excellent starting point for keeping the Earth warm enough to be habitable. Realistically any far-away planet must be able to do the same thing because space really is a hostile place and any planet that is to be habitable must have a very particular surface temperature. From the thousands of possibilities will another hot-interior, heat-losing planet ever be found? The heat being lost by planets within the solar system can readily be measured. Therefore of the factors that must be checked in order to determine if some other planet in the universe is habitable is whether or not it is losing heat. If it is, this would indicate a warm interior and provide a positive indicator for possible habitability.

While an appropriately-thin crust and a warm interior provide an excellent baseline condition for habitability, it can readily be seen that this is a fleeting condition. A warm interior means that the interior is losing heat. Losing heat means that the crust will gradually solidify and increase in thickness. The loss of heat from the interior is quite acceptable because this is the heat that is modifying surface temperature and providing a major contribution to having an acceptable surface temperature. However, loss of heat means that the crust will gradually increase in thickness. This will be accompanied by a decrease in the heat available to help maintain surface temperature. Anyone who believes that the Earth has been here in a habitable state for millions of years and will continue in a habitable state for many millions more must be prepared to explain how habitability can be maintained when the crust is several times as thick as it is now and the heat available from the interior is only a small fraction of what it is now. What source of heat will be introduced to offset the reduction from the interior? This 'window of life' is not very wide. When the surface temperature drops by (less than) 10F large areas of the Earth will become covered by snow, ice and hoar frost. This will cause the surface temperature to plummet to well below habitability. This is, in fact, what would happen if the other temperature control factors did not offset it. However, since the CO_2 factor is currently increasing on a much shorter time scale, an increasing crust thickness will probably never be noticed.

It is appropriate to recognize that there appears to be a range of CO_2 in the atmosphere that enables temperature stability. Above this range (i.e. where we are now) stability is lost. This means that long-term stability will be facilitated by an appropriate level of CO_2. (i.e. The state of the world pre-industrial) In this case heat from the interior would be a significant factor in the maintenance of habitability because with a lower level of atmospheric CO_2 with its accompanying lower level of heat retention, the heat from the interior would have been comparatively more influential.

The warm interior of the Earth is an important factor in the Earth's habitability. Will a far-away planet ever be found that has an appropriately-warm interior accompanied by an appropriately-thin crust? Or will some other heat source be identified which provides this important function? Temperature control within a very particular range is a very important factor for habitability. How will this be accomplished on a far-away planet?

6.3.2   Axial Tilt

The Earth has an axial tilt. As the Earth orbits around the Sun its orbit describes a plane. As it travels along its orbit, its axis of rotation (i.e. a line through both the north and south geological poles) points almost straight up (towards the North Star). Almost, that is, but not quite. It is tipped over a little bit. The amount that it is tipped over is called the Axial Tilt. From an observer on the surface of the Earth, the Sun rises upwards from the horizon a little differently each day as the year goes by. During the summer the Sun is quite high in the sky. During the winter it is quite low. During the spring and fall it is someplace in between. In fact at Vernal Equinox and Autumnal Equinox it is exactly above the equator. The amount of the tip-over (i.e. the axial tilt) is 23.4 degrees and is a very important factor in the habitability of the Earth.

The Moon both causes and maintains the Earth's Axial Tilt. If there wasn't any Moon, the Earth, while it might have an axial tilt, would not have a stable one. (Without the Moon) 'The Earth would tilt as much as 85 degrees off vertical. A tilt this large would be catastrophic because the seasons would not occur.' (137) 'The Moon's role in stabilizing Earth's Axial Tilt... is part of a large suite of evidences that show our home planet was designed for life.' (186) Mars also has an axial tilt but it is understood that it is not stable and will drift away from the present arrangement. Mars does have two very small moons and it is generally thought that because these moons are so small, they will have no affect on Mar's axial tilt at all (138) implying that if

The Earth's Axial Tilt

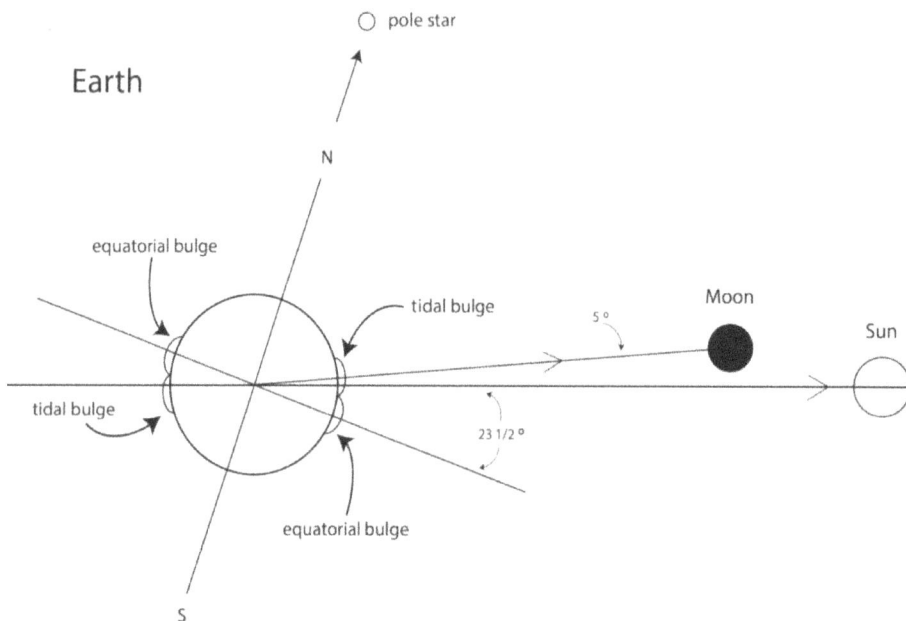

they were bigger or there was only one large one, axial tilt would be ensured. This reasoning, while popular, is not valid. While it is certainly correct that Mars requires a much larger moon with a very particular orbit to have a predictable and stable axial tilt, it is also true that Mars does not have the other factors (including an appropriate equatorial bulge (i.e. because it is too small, not fluid enough inside and does not have a large ocean)) that are required.

6.3.2.1   The Importance

The Earth's axial tilt is shown in the diagram. The axis of rotation of the Earth is tilted away from being 90 degrees to the plane of rotation around the Sun (i.e. from pointing straight up). The current axial tilt is 23.4 degrees (away from straight up) and this is the orientation of the Earth in space that gives the Earth seasons. (136) Spring, summer, fall and winter happen every year and they happen the same way every year. The seasons not only bring a variation in weather they also enable the heat from the Sun to more equitably heat the Earth. The incoming heat is spread out better. Included in this arrangement is the very particular and beneficial effect on heat retention and temperature regulation provided by the combination of the eccentricity of the Earth's orbit, the Axial Tilt, the Greenhouse Effect and the ocean. All of these factors are necessary. In particular without the Axial Tilt (even if all of the other factors were present and active) temperature regulation would be lost and the Earth would not be habitable.

While the Axial Tilt gives us seasons, the Axial Tilt in combination with the Earth's slightly varying distance to the Sun results in the southern hemisphere having summer when the Earth is closest to the Sun. However the surface of the Earth in the southern hemisphere is mostly water. Ocean covers much more of the Earth's surface south of the equator than it does north of the equator. Therefore, exposing the ocean surface to the Sun when the Earth is closer to the Sun enables the ocean to absorb more heat and warm up a little more. Water is an excellent material for storing heat and is able to readily absorb it without resulting in large increases in temperature. It stores heat much better than solid materials like soil and rock. It is most beneficial for the Earth to store heat during one part of the year because later in the year when a little extra heat could be used, this stored heat is available thereby contributing to spreading out the heat from the Sun more equitably over the whole year.

The Axial Tilt together with the varying distance to the Sun also means that when the northern hemisphere is having summer, the Earth will be a little further from the Sun and will not tend to over-heat as much (i.e. because there is more land in the northern hemisphere and land heats up more readily than water). Similarly, when the northern hemisphere is having winter, the Earth is closer to the Sun and heat input is slightly greater than it would be with any other arrangement. All of these factors working together contribute to keeping the temperature at the surface of the Earth in the best possible range. While this arrangement works to our optimal benefit, upsetting it will work to our peril! It must not be upset. Unfortunately it is being upset!

While the axial tilt enables the Earth to have seasons, seasons are not just enjoyable, they are necessary. If there wasn't any axial tilt and the Sun shone directly down on the equator all year, the equatorial regions would over-heat. In some parts of the world the temperature is currently near the limit of what human beings can tolerate. Therefore if the Sun just kept shining straight down all year on the equator, both plant and animal life would be seriously jeopardized. At the same time areas remote from the equator including the mid-latitudes and higher would be too cold. Winters in the mid-latitudes would be just as severe as they currently are north of the Arctic Circle. South of the mid-latitudes spring would be late and fall would be early. The growing season would always be short and frequently plagued by unfavourable weather. Even as survival near the equator would be doubtful and survival above the mid-latitudes would be doubtful, survival in the remaining areas would not be much better. Hence the overall habitability of the Earth would be dramatically reduced. With respect to the search for a suitable life-supporting planet near a far-away star, habitability is the factor of primary interest and if a planet isn't habitable then it will be of no further interest with respect to the search for extraterrestrial life. Just being in the thermally-habitable zone would not guaranty habitability. While the amount of incoming heat must be adequate it must also be distributed appropriately around the planet and an axial tilt enables this to happen.

As mentioned above, any planet in the thermally-habitable zone of a Red Dwarf star would have one side continually locked towards the star. This means that heat distribution would be non-existent. Further, as the gravity of the Red Dwarf star would be distorting the planet's surface and causing lock-up, that same gravity would inhibit any possibility of a moon being present. Therefore there could never be an axial tilt!

This means that there is no hope whatsoever of survival in those narrow twilight zones of planets orbiting Red Dwarf stars for more than one reason. However, this reality does not suppress the enthusiasm for the possibility of life in such places as stated by one astronomer. 'Personally, given the ubiquity and propensity for life to flourish wherever it can, I would say, my own personal feeling is that the chances of life on this planet are 100%.' (121)

While this type of unsubstantiated enthusiasm was disappointing, it was even more disappointing when the very existence of 'this planet' was cast into doubt and then discounted altogether. 'The M dwarf GJ is believed to host four planets including one (GJ 581d) near the habitable zone that could possibly support liquid water on its surface if it is a rocky planet. The detection of another habitable-zone planet – Gj 581g – is disputed as its significance depends on the eccentricity assumed for d. Analyzing stellar activity using the Ha line we measure a stellar rotation period of 102 + or − 2 days and a correlation of Ha modulation with radial velocity. Correcting for activity greatly diminishes the signal of GJ 581d while significantly boosting the signals of the other known super-Earth planets. (This means that) GJ 581d does not exist but is an artefact of stellar activity which, when completely corrected, causes the false detection of planet g.' (162) In other words the chances of life were 100% even though the host star was a Red Dwarf and the planet itself did not exist! (So it couldn't possibly have an axial tilt.)

6.3.2.2   The Cause

There are two features of the Earth that enable the Moon to generate and maintain the Earth's Axial Tilt. The two features are; 1, the Tidal Bulge and 2, the Equatorial Bulge. (Please refer to the diagram entitled 'The Axial Tilt')

The Earth's Equatorial Bulge exists because the Earth continually rotates and because it is a flexible fluid-like body. If neither of these factors existed, there would not be an equatorial bulge. However because both of them are present, the Earth bulges somewhat at the equator as it rotates. While this bulge is not very large (i.e. it is about 19 miles further from the center of the Earth to the surface of the ocean at the equator than it is at the North Pole (140)), it is very spread out and has no effect on activities on the surface of the Earth whatsoever. However it provides an element of non-symmetry making the pull of the Moon unequal over the entire sphere of the Earth. The Moon's pull peaks about 28 degrees from the equator. (i.e. the axial tilt plus about 5 degrees) This means that the pull of the Moon on the two sides of the Earth will not be equal because on one side (i.e. the lower side in the diagram) the bulge is closer to the Moon than it is on the other side (i.e. the upper side in the diagram). If there wasn't any counter-balancing force, this inequity of pull would cause the Earth to become more upright reducing the Axial Tilt until it disappeared.

A similar situation exists between the Earth and the Moon with respect to the reality that one side of the Moon always faces the Earth. This is caused by the center of gravity of the Moon being offset from its geological center. This enables the gravity of the Earth to differentially pull on the material of the Moon and keep one side facing the Earth all the time. In effect this causes the Moon to rotate on its axis only once for every circuit it makes around the Earth. As the Moon orbits the Earth, the side that is facing the Earth tries to just keep moving straight ahead without rotating but as soon as the misalignment reaches a few degrees the pull of the Earth causes the near side to once again swing around and become aligned with the Earth. This drifting and correcting goes on repeatedly. The Moon is consequently said to librate. (188)This means that it rocks back and forth slightly so that technically more than 180 degrees of its surface is visible to us from the Earth. The libration will continue indefinitely because there isn't any damping factor to reduce it. Similarly the gravity of the Moon tugs on the part of the Earth's Equatorial Bulge that is closest to it a little more than it tugs on the part on the Earth's far side. Since there is an angle of inclination in place and the Equatorial Bulge is below a line drawn from the Earth to the Moon, this tugging is in the direction of trying to pull the Earth back upright to reduce the angle. The Equatorial Bulge provides the material that enables the Moon to reduce the Axial Tilt. The Tidal Bulge does the opposite.

The gravity of the Moon creates the Tidal Bulge. Once again the flexibility of the Earth's crust comes into practical use. The Moon pulls directly on the crust of the Earth and because it is slightly flexible it will respond by rising up slightly towards the Moon. While this bulge is not extremely large it is quite well defined and has the shape of a mound almost directly in line with the Moon. Since the Earth is continuously turning, the mound reaches maximum elevation slightly after the location that is directly under the Moon. The tidal mound (i.e. bulge) therefore leads the Moon slightly and the Tidal Bulge keeps circling the Earth continuously. While there is one mound almost directly under the Moon there is also another one on the opposite side of the Earth.

The second component of the Tidal Bulge is the water in the ocean. This is the most familiar portion of the bulge and is well known to all mariners. The water bulge (or water tide) is quite measureable but it is also localized to a region almost directly under the Moon. The combination of crustal bulge and water bulge are the elements of the Earth that the Moon creates and then pulls on and in this case the pulling is towards increasing the Axial Tilt. It is generating an overturning force which, if left unchecked, would continue to increase the Axial Tilt until the North Pole was facing the Sun. In other words it would be a disaster. Without the compensating Equatorial Bulge, the Tidal Bulge would tip the Earth over and then spiral from the mid-latitudes northward until it reached the high latitudes and its torque-generating, Earth-overturning ability was no longer effective.

Most importantly, The Tidal Bulge is the second (necessary) factor required to establish and maintain the Axial Tilt. It is the gravitational pull of the Moon on the Tidal Bulge together with its pull on the 'upper' portion of the Equatorial Bulge that offsets and balances the gravitational pull of the Moon on the 'lower' portion of the Equatorial Bulge. (i.e. upper and lower in the diagram) These two forces are balanced as long as the Earth continues to rotate at its present speed. An optimal axial tilt is thereby maintained.

Referring again to the diagram entitled 'The Earth's Axial Tilt' we see that there are three bulges identified. In reality the Equatorial Bulge is continuous around the Earth but from a gravitational viewpoint there are two regions of the bulge that are influential with respect to axial tilt. The gravitational force exerted by the Moon on the Earth is dependent on distance. This is most evident in discussions of the tidal effect in general and when tides are discussed a tidal bulge towards the Moon on the side of the Earth closest to the Moon is always shown with another bulge away from the Moon on the other side. The reasoning is that the side closer to the Moon is being pulled a little harder and rises (because it is flexible) towards the Moon. At the same time the material of the Earth on the far side of the Earth from the Moon is farther away from the Moon and is being pulled with less force so it sags away from the Moon. This is the reason that tides come regularly on a twelve hour schedule. While one of the Tidal Bulges will be (nearly) right under the Moon, on the other side of the Earth (i.e. 12 hours away) there will be a second bulge. (Our discussion actually involves both bulges.)

While the Moon is continually pulling on the material of the Earth and pulling it directly towards itself, the relatively-rapid rotation of the Earth combined with the Earth rotating in the same direction as the Moon orbits, causes the Tidal Bulge to peak slightly ahead of the Moon. This effect pulls the Moon forward in its orbit. However the Earth is a sphere. This means that as it rotates (and because it is slightly tipped over (i.e. by the amount of the Axial Tilt)) the Bulge not only gets ahead of the Moon it also gets higher than the Moon. Looking at the Earth from the side as the diagram depicts, a Tidal Bulge that rises centered on some particular latitude will stay on this latitude as the Earth rotates and will therefore become higher than the Moon. Therefore the effect of the spherical shape of the rotating Earth is to pull the Moon upwards into a higher orbital plane. The Moon is not only being pulled forward it is also being pulled higher. The cost to the Earth of the Moon being pulled forward is a very slowly decreasing rotational speed and the cost to the Earth of the Moon being pulled higher is a bulge of material that is slightly offset from being in a straight line to the Moon.

Currently the orbital plane of the Moon around the Earth is slightly more than 5 degrees above the orbital plane of the Earth around the Sun. This is shown in the diagram. As mentioned, the effect of the gravitational pull of the Moon on the Earth is to provide a third and fourth bulge (i.e. the tidal bulges). One is on the Moon side of the Earth and the other is on the opposite side. 'The tides are caused ... by the effect of the Moon's gravity pulling on the Earth. ... The result is the periodic rise and fall of the Earth's major bodies of water – as well as a similar rise and fall of the Earth's bulk.' (141) The gravitational effect of the Moon on these tidal bulges together with the gravitational pull of the Moon on the 'upper' portion of the Equatorial Bulge exactly offsets the pull of the Moon on the 'lower' portion of the Equatorial Bulge. The Axial Tilt of the Earth is thereby determined as well as stabilized against minor disturbances that come our way from time to time.

In fact, it is necessary that the Axial Tilt be stabilized for the same reason that the steering wheel of a car must be stabilized. One cannot simply set the steering wheel and expect the car to go straight down the road. Minor bumps in the pavement, passing vehicles and wind all contribute to veering the car away from the intended direction. So the steering wheel must be stabilized. Neither can any random upsets in the gravitational field of the Solar System be allowed to modify our Axial Tilt.

To summarize, the Earth's Axial Tilt is determined by the pull of the Moon on two of the Earth's bulges, the Equatorial Bulge and the Tidal Bulge(s). If either of these bulges undergoes a change in magnitude, the angle of the Axial Tilt will also change. Unfortunately the Equatorial Bulge is changing (relatively rapidly) and the change will not be to the benefit of the Earth.

The Axial Tilt is absolutely necessary for the habitability of the Earth. However it will have a limited life-time because the rotation of the Earth is slowing down. If a far-away planet is being considered for habitability it must also have an axial tilt which must be operational during the period of time that the planet would be assessed for habitability. What is the possibility that such a situation will ever be identified?

6.3.2.3 Planet Mass Stability

While the mass of any sun must have long-term stability in order to be considered as a possible host heat-source for a life-bearing planet, the planet must also have long-term mass stability to ensure that the pull of its moon on its tidal bulge does not change. If, for example, the Earth became heavier, the Moon would be pulled in closer and the tidal bulge would get bigger. This would cause the Earth to 'tip over' and become uninhabitable. Unfortunately, large objects like stars and planets have a very strong gravity fields. Strong gravity fields attract matter. Over the short term this would not be a problem but over the long term it can plainly be seen that it would be a problem. Very simply, if a star was constantly getting heavier, the orbit of all of its planets would not be stable and repeatable but would constantly be getting smaller as they spiralled inward. Similarly, if a planet was getting heavier (i.e. acquiring more mass) its moon (because the moon's orbit would be dependent on the planets mass (i.e. weight)) would steadily get closer to it. The Earth absolutely must have long-term mass stability or every other factor that depended on that mass would be modified. In particular the pull on the Moon would keep increasing and the Moon would gradually spiral into the Earth. All stars and planets in the universe are also probably getting heavier as they attract more and more material from their region of space and hence they do not have any long-term stability either.

Unfortunately for those with a long-term viewpoint, the mass of the Earth is continually increasing (107) and this indicates that the Moon could never have had a stable orbit at its present distance for an extended period of time (i.e. billions of years). (If this is what is happening on the Earth, why would we expect anything different to be happening on a far-away planet?) This in turn indicates that every other factor dependent on lunar distance would have been continually changing over the millennia and will continue to change. In particular, in the distant past, the Moon would have been further away with the result that tidal magnitude would have been reduced. Also in the past the rotational speed of the Earth would have been greater. The greater rotational speed would have produced a greater equatorial bulge while the tidal bulge would have been reduced. From this combination of factors it is clear that there would not have been any axial tilt and the Earth would not have been habitable. From current observations long-term planet mass stability is not a realistic expectation which means that it is not realistic to expect long-term planet habitability either.

6.3.2.4   Moon Size and Distance

While we have really been looking for a planet all along, it is now plainly obvious that looking for a planet prior to identifying a suitable star would have been a waste of time. Similarly, trying to identify a planet which does not have a suitable moon at the proper distance would also be a waste of time. The three factors cannot be separated. In fact if neither the sun nor the moon associated with the candidate planet are not appropriate, such a planet must be abandoned as a possible candidate for life support without any further investigation. The reason is that it is absolutely necessary that the candidate planet have an appropriate moon or else it will not have an appropriate axial tilt.

The Moon, even though it is much smaller and lighter than either the Sun or the Earth, has significant gravity and therefore continually attracts material from space into itself. The evidence that this has been happening over the years is plainly seen by looking at the Moon's surface. Numerous objects have struck the Moon and left marks identifying where they impacted. The larger impact marks have been counted. There are at least 200,000 impact marks on the Moon (106). More impacts happen continually. We have the major impact of 1178 (189) as well as the ongoing recordings of impacts (148) as evidence that this is happening. The Moon also has mass concentrations (i.e. mascons) beneath the maria. These are suggestive of large objects having broken through the surface and displacing molten interior material which proceeded to ooze out onto the surface. The impacting objects then stalled before sinking all of the way to the center.

If the Moon is getting heavier and heavier, the force it exerts on the Earth will become greater and greater. This means that the force it exerts on both the tidal bulge and the equatorial bulge will be continually increasing. The effect on the tidal bulge is compounded because the increased pull will generate a larger tidal bulge as well as pull harder on it. The size of the equatorial bulge would not be affected but it would be pulled a little harder. The balance between the two would be upset. Clearly, only one size of Moon will do and it must have long-term mass stability or the Axial Tilt of the Earth could not possibly have long-term stability either.

If long-term lunar mass stability is necessary for the Earth's habitability, why would we expect anything different for a far away planet set-up?

6.3.2.5 Magnitude

It is the Earth's axial tilt that enables so much of the Earth to be habitable. The magnitude of the axial tilt is currently 23.4 degrees (147). (Over the years this value varies slightly due to the slightly elliptical orbits of the Earth and the Moon.) This is, in fact, the optimum angle enabling the habitable area of the Earth to be maximized. With any other angle, less area would be hospitable to life and the overall habitability of the Earth would be reduced.

6.3.2.6 Maintenance

Due to the inertia of the rotating Earth, the axial tilt will have considerable stability against minor upsetting forces that might come our way from time to time. However it is questionable if stability could be maintained if a major disturbance should occur. For example, in 1979 a giant comet (Howard-Koomen-Michels 1979X1) came into the inner solar system and went right into the Sun. (101) It came in from the far side of the Sun which was a great relief for Earth dwellers because it was so large it would have disturbed our orbit as well as our angle of inclination. The head of that comet was larger than the Earth which, for a comet, is unusually large. (101) If a comet that size even passed within a million miles of the Earth we would be in danger.

With a solar system that is so full of surprises it is a wonder that the Earth hasn't suffered some life-terminating disturbance (like an asteroid shower) already. After all, objects like comets and asteroids are continually moving through the solar system and they certainly have no regard for life on Earth. What would the situation be like in far-away solar systems? Could a planet in a far- away solar system avoid life-threatening disturbances long enough to enable the complex circumstances required for habitability to be set up?

Mars has two moons but neither of them is very large and neither of them is effective at generating and maintaining an axial tilt. Therefore the axial tilt of Mars will wonder and is expected to drift until one of the Martian poles temporarily faces the Sun. (149) Any similar arrangement for the Earth would be a disaster. None of the other planets in the solar system have any better arrangement. In fact none of the other planets in the Solar System have any moons that are able to provide an axial tilt function at all.

While the Earth's axial tilt is very close to optimum, it is transient and does not have long term stability. Why should we expect anything different from a far away planet?

6.3.2.7 The Current Situation

The Earth is losing rotational energy. In fact, whenever movement, friction or heat is involved, energy is always lost. Heat is a form of energy but heat cannot be stored or retained. It always escapes into space and is lost forever. Whenever some form of energy is converted to heat energy that entire system will be running down and sooner or later come to a halt. The rotating Earth is no different. The Moon raises the tides and the Earth turns, causing the Tidal Bulges to travel around the Earth continually. 'The movement of the tidal water ... produces enormous amounts of kinetic energy (i.e. energy of motion) ... which ends up as heat. The tide flow moves forward in the ocean area largely unhindered until it runs into the continents, resulting in water crashing up against the continental borders. Consequently the coast line runs into a mass of water at every high tide, a collision that causes the Earth to lose rotational energy.' (150)

Tides have been recognized as a major source of energy for some time and tapping into this energy source would provide enormous quantities of 'clean' non-carbon, non-radioactive energy. One proposed project is in Nova Scotia, Canada. The Bay of Fundy is located between Nova Scotia and New Brunswick and boasts the 'world's highest tides'. (142) These tides embody vast amounts of energy to the degree that if only a small portion of it was captured at the upper end of the Bay, a 'capacity of 4,864 megawatts ((which) would be the equivalent of seven conventional Candu nuclear power plants) would be available.' (151) This, however, is a very small fraction of the energy that is being dissipated by the tides around the world every day. 'The masses of water affected by the tidal water movement are enormous, as will be clear from a couple of examples. Into one small bay on the east coast of North America – Passamaquaddy – 2 billion tons of water is carried by the tidal currents twice each day; (and) into the Bay of Fundy, 100 billion tons.' (143) It is also fitting to recognize the situation in Northern Canada. Hudson Straight (named after Henry Hudson) connects Hudson Bay to the Atlantic Ocean. Tidal water flows back and forth through Hudson Straight twice every day. The straight is more than 100 miles wide and the flow is vigorous both ways. Early explorers described the current thus; 'The currents were so strong as to turn a ship around.' (157) Further, they reported that 'the Flood Tide ran three hours to the Ebb Tides one.' (157) On another occasion the sailors reported that there was 'a westward running current' making them think they were 'hot on the trail to China.' (158) If water is flowing vigorously through a waterway that is 100 miles wide, one can only wonder just how much water is involved. All of this water is flowing into and out of Hudson Bay and James Bay. Even down at the south end of James Bay, several hundred miles from Hudson Straight, the water level changes by several feet from low tide to high tide.

Many years ago some early explorers sailing the Straight reported that an island more than 20 feet high sank as they were watching it. Obviously the island did not sink. The water level simply increased. The amount of energy involved in such a situation truly is enormous and all of it develops at the expense of the rotating Earth.

Another couple of examples would seem to be in order. The Saint Lawrence River flows eastward and drains the Great Lakes which form a considerable portion of the Canada-USA border. The river/estuary varies in width from about 70 miles wide at Sept Iles (near the mouth) to about 10 miles wide just below Quebec City. At Quebec City the elevation change in the water level from low tide to high tide is between 15 and 20 feet. The distance between Quebec City and Sept Iles is more than 350 miles. As with all tidal situations this change takes place twice every day. How much energy is involved in moving this massive amount of water and where does that energy come from?

China boosts the most spectacular Tidal Bore in the world and people come from around the world to see it. (184) (Tidal Bore is the name given to a tidal water flow that is coming in so fast that the leading edge of the flow forms a crashing, turbulent wall of water followed by a flowing sheet of water as deep as the leading edge.)With the Chinese tidal bore, the flow can be up to 35 feet high. The flow of water following the leading edge moves so fast that a person on foot would not be able to keep up with it. Even a cyclist would have to work to keep ahead of it as it flows into the narrower parts of the estuary. The estuary where this happens is several miles wide and as with all tidal movement, it occurs twice every day. The energy required to move all of this water truly staggers the imagination.

While these examples involve a lot of energy, similar situations are happening all around the world every day, year after year. The amount of energy being dissipated is very great and its loss is causing the Earth's rotation to slow down!

6.3.2.8   The Earth's Slowdown

Time, its measurement and recording, has been a preoccupation of mankind for thousands of years. People have devised numerous ways of identifying how much time was involved for all kinds of events from the tracking of the seasons to the length of the day, Time measurement devices have continually become more accurate and reliable as well as more versatile. For example, a major advance in timekeeping was achieved when a clock capable of operating on board a heaving ship was devised. This enabled measurement of longitude as the ship sailed across the ocean. Measurement of latitude preceded this development but with both measurements available, sailors could always know where they were on the surface of the Earth. With this development navigation technology was dramatically advanced.

For some considerable time the length of a day has been used to keep track of time and enable clocks to be reset. A day was always 24 hours long so every day clocks could be reset and synchronized. If the length of a day varied by a second or two there would be no way of knowing because such a short amount of time would not have been detectable. The situation changed dramatically in 1972 when the first Atomic Clock was introduced. (146) This provided a much more precise way to keep track of time – in particular the length of a day. Seemingly up to that time there was little awareness that the length of the day could actually be changing. Of course many things change in nature and as long as the changes are not dramatic they might not be noticed. However, if the length of the day was changing relatively rapidly, it would be cause for concern.

The Atomic Clock is continually measuring the length of the day and indicates that the Earth is rapidly slowing down. Days are getting longer all the time and to keep the clock in sync with the actual length of the day an additional amount of time is inserted into the clock on a regular basis. This insertion is called a leap-second and one leap-second is added to the clock about every 500 days. (145) Since the clock was first introduced, 29 leap-seconds have been added. The length of a day is therefore slowly but continually increasing indicating that the days on Earth are slowly getting longer all the time.

 This observation has been cross-checked by observing solar eclipses. 'Historical data on the locations of eclipses have allowed us to determine the rate at which the Earth's rotation is slowing because of tidal braking.' (144)

The energy to move the tidal water is continually being lost (as heat) at the expense of the rotation of the Earth. (Please refer to the diagram entitled; The Lengthening Day) Or to put it another way, the rotational energy of the Earth is powering tidal water movement and this causes tidal energy to be available. However, with all of the tidal water moving around, energy is continually being lost. Since this energy is provided by the rotation of the Earth, the Earth's rotation is slowing down and it will continue to slow down as long as ocean tides exist.

As the Earth slows down the equatorial bulge becomes smaller. The Moon will continue pulling on the Tidal Bulge but the balance between the equatorial bulge and the tidal bulge will be lost. The Axial Tilt will increase and the Earth will 'tip over' within another few hundreds of thousands of years. The Axial Tilt is necessary for life on the Earth but it will be a relatively short-lived temporal 'window of life'. Could a similar 'window of life' exist elsewhere?

When the change in the Axial Tilt starts to happen it will proceed quickly. The result will be a dramatic increase in the Axial Tilt well within another few hundreds of thousands of years. The increase will make the Earth uninhabitable well within the time required for the rotation rate of the Earth to slow down enough for a day to be twice as long as it is now. This is not good news. In fact, the obvious non-linearities involved indicate that the Earth could 'tip over' within a fraction of the million years or so before the day becomes twice as long and very likely within a small fraction of that much time. When the Earth does tip over, it will become uninhabitable and will remain so from then onwards.

With similar circumstances it would be reasonable to expect that something similar would happen on otherwise-habitable planets elsewhere.

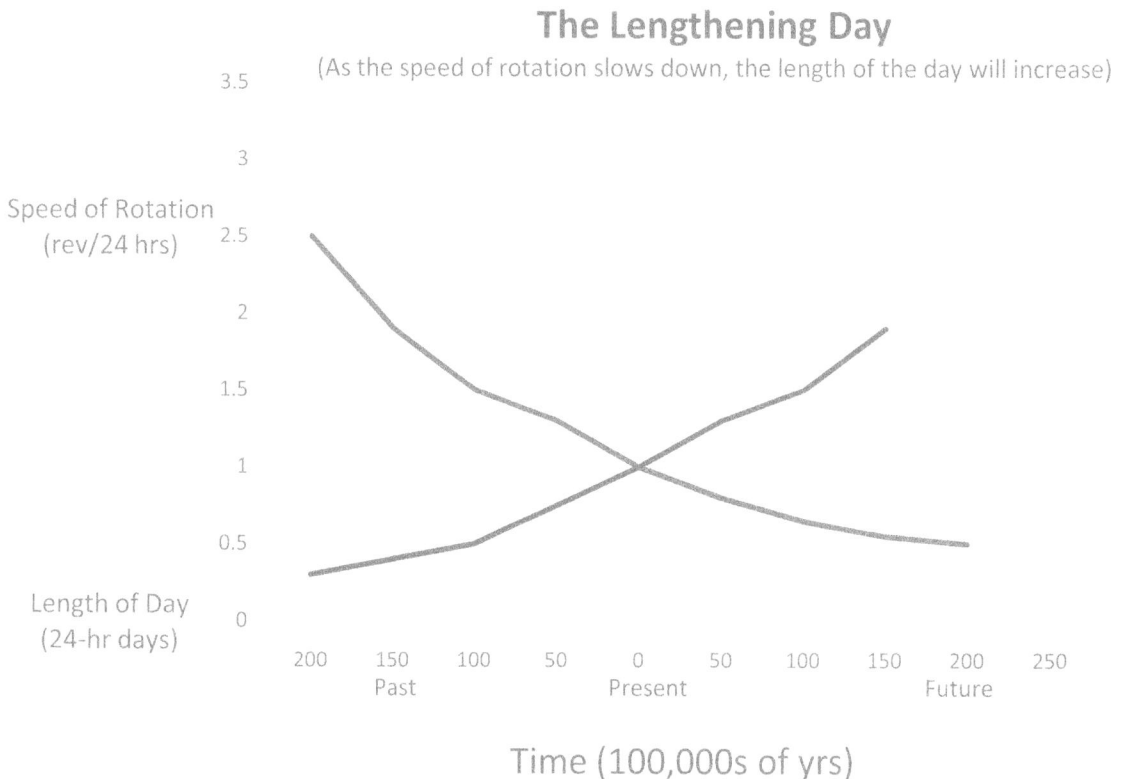

The Lengthening Day

(As the speed of rotation slows down, the length of the day will increase)

Speed of Rotation (rev/24 hrs)

Length of Day (24-hr days)

Time (100,000s of yrs)

(This diagram is intended for discussion purposes only and is not intended to accurately predict the future situation.)

6.3.2.9   Comment

What is one to make of this? While we have obviously been discussing the Earth, it is necessary that something very similar be happening on any far away planet that is suspected of being habitable. Getting a planet set up for habitability is not a trivial affair. Temperature control in exactly the right range is a minimum requirement and the spreading out of the incoming heat is fundamental to temperature control. On Earth the Axial Tilt plays a major part in achieving the right setup so the discussion temporarily shifts to how an axial tilt could be arranged. Numerous factors are involved. Could this be happening elsewhere? Why would anyone expect an arrangement similar to the one on Earth to be happening elsewhere? While the ocean tides have several beneficial effects, because the tidal water moves, energy is lost. Whenever energy is lost, movement cannot continue forever and will slow down. This is just basic physics and expectedly basic physics is acting throughout the galaxy the same as it does on Earth. Therefore far-away planets must also be dealing with basic physics and continual energy loss so whatever arrangement they currently have will not continue either but will peter out as time goes by just as it will on the Earth. And the time involved is very short. This in turn means that any window of discovery must also be short. Even if it was happening, any explorers would have to arrive within the 'window of habitability' or they might as well not come at all. The greater question remains; is it happening? Why would we expect that a set of circumstances similar to what is required for an axial tilt on Earth to appear on some far-away planet, continue for a few thousand years (allowing for discovery) and then peter out just as they will on Earth?

6.3.3   The Greenhouse Effect.

The Earth enjoys a Greenhouse Effect. While everyone appreciates that a greenhouse can be a suitable location to grow plants in adverse weather, many do not understand the reason. It is simply because the cover on the greenhouse helps to keep the heat inside. Plants can grow when the temperature is reasonable but they probably will not grow if exposed to the chilly conditions that develop outside. So the most important part of the greenhouse is the cover. Similarly an important factor in retaining heat and making it possible for the Earth to be habitable is an atmosphere which includes a Greenhouse Effect (i.e. a heat retention effect).

The atmosphere of the Earth mostly consists of oxygen and nitrogen. In fact, these two gases constitute more than 98% of the atmosphere. However from a heat retention viewpoint they do not provide any help at all. Heat retention is achieved by the small remaining portion. (i.e. less than 2%)

The three most important factors in providing the Greenhouse Effect are; a, CO_2 (carbon dioxide) b, water vapor and c, clouds. Carbon dioxide is usually called a driving factor. (128) This is because it isn't modified by any other factor such as heat. Carbon dioxide provides its heat retention characteristic simply dependant on its concentration. More CO_2 will provide more heat retention. Less CO_2 will provide less heat retention. The same cannot be said for water vapor. While CO_2 provides approximately 20% of the Greenhouse Effect, water vapor provides about 50%. This is not really very good news. If the temperature should be forced up a few degrees by CO_2, the atmosphere on the average will be able to retain more water vapor. (129) This increase in water vapor will cause the temperature to increase even further. This can become a viscous cycle where the hotter it gets, the hotter it gets.(A possible termination to ongoing temperature increases could occur if the extra water vapor resulted in increased cloud cover to the point where the increased cloud cover reflected more heat away. Therefore a terminal condition for an increase in temperature might occur when cloud cover reached some critical unknown level.)

The CO_2 inventory has increased by more than 40% since 'preindustrial times' to almost 400 ppm (i.e. parts per million) today. (130) This, undoubtedly, will have been accompanied by an increase in the amount of water vapor in the atmosphere with the net result that the average surface temperature around the world has increased. While it is hoped that future increases can be held to 2C, even that much further increase in temperature is being viewed as a disaster. Unfortunately with the melting of the permafrost in the Russian bogs in 2006 and the accompanying outpouring of both CO_2 and methane (overwhelming man-made CO_2 increases) there is no hope whatsoever that ongoing temperature increases can be avoided. (202)

The surface temperature of the Earth absolutely must be regulated (i.e. controlled) to within a very narrow range or the Earth would be in danger of either over-heating or over-cooling. Either result would terminate life on the Earth. The temperature that the Earth requires for animal life to survive and plants to grow is very particular and it must be held right where it is presently found. In fact, at the present time, the temperature at the surface of the Earth is measured to be increasing. This is understood to be the result of an increase in one of the main greenhouse gases. CO_2 is increasing in the atmosphere and while it is only the third-most influential of the greenhouse factors (after water vapour and clouds (135) see Table One) the increase in temperature is being viewed as an impending disaster. In fact, it is the fervent hope of many scientists and world leaders that if (as mentioned) the increase can be held to 2C the Earth might survive. Such a small amount of temperature increase seems very insignificant and actually it is insignificant. However it is much more dangerous than it would casually appear because of the secondary effects that will develop. (i.e. an increase in water vapor)

In reality, the actual portion of greenhouse effect that each factor contributes varies depending on which factors are actually available at any particular time and place. For example, in the Sahara Desert at night there is very little water vapor in the air. Therefore the other factors will represent a much greater fraction of the greenhouse function but the overall effect will still be reduced and the temperature will drop.

Table One - The Greenhouse Factors

The Factor	Portion of Greenhouse Effect (%)
Water Vapor	50
Clouds	25
CO2	20
Methane	1

While the importance of the Greenhouse Effect for habitability can plainly be seen with respect to the Earth, it is hard to imagine how any other planet could be habitable without one. Temperature control is a difficult criterion to meet. Could it be achieved without a Greenhouse Effect? Could a far-away planet enjoy temperature control without having a host of factors similar to Earth's in place? Even on the Earth with an active Greenhouse Effect, places such as the Sahara Desert in Africa become cold when the Sun goes down. The CO2 factor will be present in the desert night air but due to the dryness, water vapor will not be present. Therefore at night it becomes cold. This happens even though the Sahara Desert is relatively close to the equator and it means that the Sahara Desert is not really habitable because with even occasional frost, crops could never be grown. All of this is happening even though the planet is well within the habitable zone of the Sun! What would happen on a far-away planet? Would it freeze every night? How then would food plants be grown? How would any plants grow? Clearly, if a far-away planet became too chilly at night it would not be habitable even though the average surface temperature might be well within the habitable range and the planet might be within the habitable zone of its star. Therefore it is plain to be seen that it would take very little deviation from a set-up like the one on Earth before a useful Greenhouse Effect on a far-away planet would simply not exist.

6.3.4   The Ocean

The Earth has a very large ocean which is a major factor in the temperature control of the Earth. The temperature over the surface of the Earth cannot be allowed to wander very far, especially in the downward direction (because of the impossibility of recovery). When we consider the range of temperature that is possible in this universe we can appreciate that the temperature variation from our current near-optimum level of 59F (15C) (166) to freezing is really a very small fraction of the temperatures that are possible. If freezing ever occurred over any significant fraction of the Earth's surface there would be no recovery. (160) The incoming heat from the Sun would be reflected away from the Earth by the snow and ice and there simply would not be enough heat from the Sun left to raise the temperature. (167) Heat that is reflected away can never be recovered. This is problem number one. Problem number two follows immediately and relates to the inability of air to retain water vapour when it cools down. Cold air cannot hold very much water vapour. We recall that water vapour is the most influential greenhouse factor (135) and without water vapour in the air the temperature would spiral down - even during summer near the Equator. (Remember the Sahara Desert which occasionally freezes at night even though it is close to the Equator.) It would only stop going down when any further drop in water vapour level did not result in any further drop in temperature. Unfortunately this would not happen until the temperature had dropped to well below freezing. From such a state there would be no chance of recovery.

The ocean plays a major role in temperature stability in several ways. First of all we recognize that the ocean is very large. 'Earth is a little more than two-thirds covered by ocean and a little less than one-third land with the ocean on average about 4 km. deep (with a) pressure at the bottom ..... about 6,000 lbs/square inch' (161) 'About two-thirds of Earth's land area is in the Northern Hemisphere so that about 57% of the ocean is in the Southern Hemisphere, 43% in the northern; the Northern Hemisphere itself is 61% ocean and the Southern Hemisphere is about 80% ocean. The volume of the ocean is about $1.3 \times 10(18)$ m(3). In other words the ocean is very large and the dominant factor on the Earth's surface. All of this is significant with respect to smoothing out variations in temperature at the surface of the Earth. The ocean simply soaks up heat when it is available and releases it again when the temperature of the atmosphere drops. In particular, during summer in the southern hemisphere with the Earth a little closer to the Sun there is a greater amount of heat available. The ocean takes up some of this heat without any significant change in temperature. Then a few months later during winter in the Southern Hemisphere this heat is released back into the atmosphere.

'The ocean moderates the climate by taking in heat when the overlying atmosphere is hot, storing that energy and releasing it when the atmosphere is cold.' (169) One particular example (i.e. New York and San Francisco) will illustrate this effect. 'The two cities have similar latitudes and both are on the coast yet the range of temperature is enormously larger in New York. The highs are higher; the lows are lower. The difference is because the climate of San Francisco is maritime while the climate of New York is continental. The winds that blow over New York come from the west so they blow over land. The winds that blow over San Francisco have come from the ocean.' (173)From this one example we can readily see that with such a huge ocean available (on a global scale) there will be a climate moderating effect.

The ocean transports heat pole-ward in both hemispheres. The transport is associated with a release of heat into the atmosphere at high latitudes whereas the ocean is being heated by the atmosphere at low latitudes (as well as by the Sun shining directly on the ocean at the low latitudes). As an example, heat transport happens in the North Atlantic because of the Gulf Stream which flows north-east from the vicinity of Florida all the way up to the vicinity of Spitzbergen Island which is much further north than Iceland. All along its pathway the climate is moderated. (171) In particular, consider the Faroe Islands which lie directly in the path of the

Gulf Stream (between Iceland and Norway). On the Faroe Islands the temperature variation over the entire year is only from 37F to 54F. (170)

This type of heat transport is repeated numerous times in many other parts of the ocean resulting in a considerable overall climate moderating effect for the Earth. In other words without heat transport by ocean currents the Earth would suffer much greater temperature variation over its surface than one might expect since we are right in the middle of the thermally-habitable zone of the Sun. What then might one expect on a far-away planet which might also be right in the middle of the thermally-habitable zone of its star? With a large ocean such a planet might have a chance of being habitable but without a large ocean the prospects for habitability would be much reduced.

6.3.5   Rotation

There are two factors affecting the Earth's rotation. The energy lost by tidal water movement has already been discussed and is a phenomena with a relatively short time-frame. Great amounts of energy are lost as heat by the motion of the tides. This loss will affect the rotation of the Earth on a relatively short time-frame of tens of thousands of years and not millions. The second factor slowing the Earth has a much greater time-frame of millions of years and in fact tens of millions of years. Tidal action is also involved.

In this case when a line is drawn from the center of the Earth through tidal maximum and extended into space, it is found that the line is slightly ahead of a line through the center of the Moon. This is because of the Earth's rotation. The bulge of the tide, being slightly ahead of the Moon, pulls on the Moon to move it forward in its orbit. The tidal bulge is pulling the Moon forward and hence is speeding it up. The speed-up is achieved at the expense of the rotating Earth with the result that the Earth is slowing down. Thankfully this is a slow, long-term process. The ultimate result will be an Earth that is turning much slower and a Moon that is higher up and farther away. The slowing down will peter out when the Earth's rotation matches the orbital speed of the Moon. Since the Moon will be about 50% farther away by then, its orbit around the Earth will take about six weeks. This means that a day-night period on the Earth will be about six weeks long with a period of light of about three weeks and a period of darkness of about three weeks. This will be an impossible situation to deal with because during the night the surface of the Earth will cool. In fact it will cool into the freezing range winter and summer. When the Sun comes up the surface of the Earth will once again start to warm up. However since snow and hoar frost will cover everything most of the much-needed heat from the Sun will be reflected away. This means that the Earth will not completely heat up again. While there will be three weeks available, it will require most of that time to bring the temperature into the melting range. Since the dark side will be cold and the warm side will not really be warm the overall amount of water vapour in the air will be seriously depleted. The Greenhouse Effect will be lost. The Earth will become a 'Snowball Earth' and habitability will be lost. In fact it will be lost long before the Moon reaches its maximum distance away.

While this situation will never come about because of the other devastating developments it points out the importance of having appropriate periods of night and day. The 24 hour cycle works very well but a six week cycle will not work at all. Will a far-away planet ever be found that has appropriate periods of night and day?

6.4   Material State

As can be seen from the necessity for a habitable planet to have an axial tilt, the material state of such a planet must include a hot, fluid interior covered by a relatively-thin solid crust. This arrangement will enable both a tidal bulge and an equatorial bulge to form. While rotation is also necessary, rotating too fast or too slow will not be acceptable. The objective is to have a useful equatorial bulge and such a bulge would not be useful if it was either too large or too small. The required bulge would be part of the axial stability function. A proper spin rate and a fluid interior are absolutely necessary. 'I don't imagine many geophysicists, when asked to count their blessings, would include living on a planet with a molten interior, but it's a pretty near certainty that without all that magma swirling around beneath us we wouldn't be here now.' (182)

The Earth is constructed as a large fluid/semi-fluid sphere with a comparatively thin crust. The flexibility provided by this arrangement enables bulges to form if the appropriate forces are applied. If the Earth was a solid body, bulges would not be able to form. However bulges can form on the Earth due to either gravity pulling on the Earth or by the rotation of the Earth and both types of bulges are involved in establishing and maintaining the Earth's Axial Tilt. The pull of the Moon's gravity causes the tidal bulge to form. The rotation of the Earth causes the equatorial bulge to form. Clearly without a material state similar to the Earth's, the necessary bulges would never form on a far-away planet and any hope of an appropriate Axial Tilt would be lost and with it any hope of habitability.

6.5 Size and Mass

A planet can be neither too large nor too small. While there is some leeway with this factor the leeway is not really very great. The Earth is just large enough to hold an atmosphere. While it cannot hold hydrogen which raises a doubt concerning our long-term water supply, it can hold both oxygen and nitrogen without difficulty. If the diameter of the Earth was only about one-half of its present value (assuming similar density) the mass would be reduced to about 10% of its present value. The Earth's force of gravity would be reduced (to about the same as Mars) and escape velocity would be lower. It would be impossible to hold a substantial atmosphere under such conditions. On the other hand if the Earth had a diameter that was twice as large as it is at present, its mass would have increased about eight times and the force of gravity would be approximately twice as great. In this case an oxygen-nitrogen atmosphere could be retained without difficulty (as well as several much less desirable gases like ammonia) but it would be more difficult to evaporate water. For reasons like this astronomers prefer that a potentially-habitable, far-away planet be in the size range of 0.83 Earth to 1.2 Earth. Of the (approximately) 4000 extra-solar planets discovered, only five meet these criteria. (174)

6.6 Atmospheric Circulation

The Earth's atmospheric circulation patterns involve three definite loops. The one closest to the equator is called the Hadley Circulation Cell and operates as follows. The heat from the tropics causes air to rise. The rising air entrains massive amounts of water vapour and torrential rains result as the warm humid air rises and cools. This is the current pattern in the atmosphere and if the equatorial regions were even hotter more moisture would be entrained resulting in even greater downpours than we have at present. The rising air dries out, partly because it loses its load of moisture and partly because it expands as it rises away from the surface of the Earth. This cold dry air then drifts north and descends back down to the Earth's surface. As it descends, its relative humidity drops and desserts are created where it comes back to the surface. In fact, most of the major deserts of the world exist in two bands centered about 30 degrees from the equator because of this down-flow of very dry air. (183) Unfortunately, if the Earth was sitting 'upright' (i.e. without an axial tilt), this air would be descending right down onto the only remaining thermally-habitable region in existence making it also uninhabitable because it would be just too dry to grow crops.

Incidentally, the fires of California are indicative of the fact that California is becoming desert. This is not good news because the fires are destroying personal property but more importantly for the big picture because California will no longer be able to supply food to the rest of the continent. Where the food that we currently get from California will come from is very uncertain. In recognition of our basic topic, (i.e. habitability) the Earth is becoming less habitable because of these developments. The reason that this is happening is because the additional heat that the Earth is retaining, due to the increasing Greenhouse Effect, becomes entrained in the rising air (of the Hadley Cell circulation) and takes slightly longer to cool as it travels northward so it will be able to travel a little further north before it descends. Consequently, California will become increasingly desert-like during the coming years and the Earth will become a little less habitable.

Similar reasoning applies to the past. Within a few tens of thousands of years in the past the Earth would have been rotating much faster and the equatorial bulge would have been much greater totally dominating the Axial Tilt. In fact, the higher speed would have resulted in there not being any axial tilt. The Sun would have been shining straight down on the equator all of the time. The equatorial regions would have been over-heated, the regions immediately north and south would have been deserts (due to the down-washing of the Hadley Cell circulation system) and the regions north of approximately 45 degrees would have been locked in continuous winter - complete with permafrost. In between there would be a narrow region where life might have existed but minor variations in atmospheric circulation from either side would have made it very difficult.

There is a corollary to the above line of reasoning; *If the Earth had ever been rotating without an axial tilt it is inconceivable how it could ever have acquired one.*

While we can safely assume that any potentially-habitable far-away planet must have an appropriate atmosphere, it must also have a beneficial atmospheric circulation system. While this quite possibly might never be observable, it would never-the-less be necessary. Nobody could ever live on a planet that was what California appears to turning into (i.e. a desert). What is the probability that a far-away planet will have an appropriate atmospheric circulation system?

6.7   Life Support Systems

While in reality there are numerous life support systems in operation on the Earth only four will be discussed here. The four chosen are obvious. Others that might have been chosen include; The Carbon Cycle and the Nitrogen Cycle but reviewing the four included will enable the necessary insight to be gained to help everyone realize that the provision of a life-support system is a necessarily-complex affair. Just as there are no simple forms of life there is no simple life support system.

6.7.1   Potable Water

If a planet is to support any form of animal life there must be an ongoing source of potable water (i.e. drinking water). 'Ongoing' is specified because water is very easy to contaminate and there isn't any way that a quantity of water can be left standing unprotected for any significant period of time without acquiring some undesirable substance. Besides being an acceptable home for a diversity of animal life, water is also home to an even greater diversity of microscopic life some of which can cause humans to become ill.

Many different types of salts and minerals dissolve quite readily in water. While small amounts might not cause trouble, excessive amounts might be fatal. As water flows over salts or minerals on its way to the ocean, some of these materials dissolve into the water and of course some dissolve much more readily than others. The Jordan River flows from Syria through Israel to the Dead Sea. On its way it picks up salt. This makes the Dead Sea very salty to the extent that there is virtually no form of life in it. In fact, the Dead Sea salt level has long since reached saturation (i.e. about 30%) and the water simply cannot hold any more. Therefore as more is brought down by the Jordan River equivalent quantities precipitate out. The shores of the Dead

Sea are piled up with salt, some of which is commercially harvested. While this is economically advantageous, with the Dead Sea at the salt saturation level no form of animal life can exist in its waters. Nobody goes fishing in the Dead Sea. Most of the water which pours down the Jordan River has passed through the Sea of Galilee. At the Sea of Galilee the water is not salty and for a long time there has been an active fishing activity on this Sea but between there and the Dead Sea salt is picked up making the water inappropriate for either human or animal consumption.

Salt is also picked up by numerous rivers which flow into the ocean with the result that all of the oceans are salty to some degree making most of the water on the Earth undrinkable.

pH is a term used to identify the acidity or alkalinity of water. If the pH is neutral, the water is neither acidic nor basic. A pH close to neutral is necessary for habitability and is not only desirable for life on the Earth but is a question properly raised with respect to the possibility of life elsewhere. 'Even if a planet has oceans, does the water have a pH neutral enough to permit cells to grow?' (190) A pH of 7 is neutral. Lower than 7 is acidic and higher than 7 is basic. (191) If certain materials are added to water the pH level can be shifted into either of these less desirable regions. For example adding lime to water will make it more basic. In fact this has occasionally been done to offset too much acidity. Of course adding any acid to water would make it acidic and this has been unintentionally done as well.

During recent years the pH of numerous lakes has been modified by pollution and the pH levels have drifted away from neutral. In particular, many lakes within a few miles of the large smokestack at Sudbury, Ontario, have been subject to acid rain from this smokestack. As a result the water in these lakes became very clear. While 'very clear' is desirable for viewing the bottom of the lake, it is not at all desirable for the minute forms of life in these lakes. As a consequence they die and any creatures dependant on them die as well. This includes fish. If there are no fish there will be no fish-eating birds and the entire wildlife scene is negatively modified. Fortunately, improved technology and more relevant legislation have resulted in many of these lakes becoming murkier and once again being able to support a diversity of wildlife.

The pH of water is a two-sided window. A shift is possible either way from neutral but neither direction is desirable and a serious shift either way would render the water unfit as life-support material. It is always desirable to have a close-to 'neutral' pH. Occasionally a body of water is found that is too far from being neutral and therefore it cannot be used by either humans or animals including fish. Within recent history a small lake in Ontario, Canada became contaminated by an acid spill. The fish all died and it took several years before they reappeared. The acid partially dissipated by natural means and partially by having correcting material introduced. It is worth noting that the actual amount of acid that entered the lake was only a very small percentage of the lake volume. It was a tanker truck load which by comparison to the volume of the lake was only a very small percentage. Never-the-less the lake became so contaminated that it could not support any significant diversity of life.

While there is a very great amount of water on the Earth only a very small percentage of it is useful for human and animal consumption.

If a far-away planet should be discovered that has water on it what can we assume about that water's usefulness? Will it be drinkable? Anyone observing the Earth from some far-away vantage point might conclude that the Earth should be habitable because of the great amount of water that we have. However as all residents of Earth know, we cannot drink ocean water even though there is plenty of it. For drinking purposes it is polluted and when we recognize the relative ease with which water can be contaminated the possibility of finding a source of drinking water on some far away planet seems pretty remote.

6.7.2   The Magnetic Field

The Earth is a magnet. Therefore it has a magnetic field. The term 'field' implies that there is some influence remote from the object, which is causing the 'field'. The Earth also has gravity. The force of gravity is similar to a magnet because there is an effect far away from the Earth itself. While nobody has been able to explain this remote effect, the magnitude of it can be calculated. It is understood that the force of gravity between the Earth and the Moon keeps the Moon in orbit around the Earth. Similarly, there is a force of gravity between the Earth and the Sun which is recognized as keeping the Earth in orbit around the Sun. We have Isaac Newton and Albert Einstein to thank for the mathematics involved but we do not have anybody to thank yet for explaining just how the force of gravity works. The theories which have been advanced so far are very tenuous but this is not important compared to the fact that the magnitude of the forces can be calculated.

As the force of gravity acts remote from the Earth and can be accurately calculated, the magnetic field of the Earth also operates remote from the Earth and its magnitude can also be calculated. While gravity is credited with holding everything in place on the surface of the Earth, the magnetic field gets very little attention. It is handy for operating compasses. If a person is away from well-known areas, knowing which way is north will be helpful in determining which way to go. A compass and the magnetic field may therefore be helpful for our survival away from home but there is another way in which the magnetic field is helpful all the time.

The magnetic field of the Earth is quite strong. It therefore extends well out into space and will have an effect on all incoming particles, which may be influenced by a field of this nature. Unfortunately the Earth is the target for a lot of particles from space including very small particles, which are called cosmic rays. These particles are called cosmic because they come from the cosmos, which is just another word for space. It may seem like a more exotic term but space is space and these particles come to the Earth from space at very high speed. Also they are very small. They are atoms but they are not complete atoms and this is the reason that the magnetic field is so important.

Since these cosmic rays are very small and move very fast they do, in many cases, come right through the atmosphere and go right into the Earth. They could go right through a person on the way. Since these particles are atoms of various size, they might interact with the human body at the atomic level. They might crash right through the surface layers of the body and damage some part of the complicated structure which causes the cells in the body to divide. If this happens the cells will not divide properly. Tumours may result, which is a very unwelcome development.

The magnetic field of the Earth has an influence on many of these unwanted particles from space and actually traps them high above the Earth. In this way most of the particles which are rushing through space do not actually come to the surface of the Earth at all. The magnetic field has prevented them from getting here and in this way it is acting like a giant shield. It is definitely safer to live on the Earth when this shield is operating. It is also to our benefit that it is transparent so we can still see the stars and sunlight can get to the surface of the Earth.

These little particles, which would have come to Earth, become trapped high above the atmosphere in a region, which is called the Van Allen Radiation Belt. It is the magnetic field of the Earth, which causes this radiation belt to exist. If the magnetic field disappeared, the Van Allen Radiation Belt would also disappear. In such a case, the Earth would be fully exposed to all incoming cosmic radiation and the harmful effects that this would bring.

The magnetic field may therefore be considered as a "window of life" because as long as it exists, it will be much safer to live on Earth. On one end of the scale, a stronger magnetic field might offer even more protection but just how much magnetism a human being can stand is not well known. It is safe to say that it is a lot more than we have at the present time, but there would, of course, be an upper limit. This, however, does not directly concern us. We definitely benefit from the shield generated by the magnetic field and therefore the magnetic field is properly referred to as a "window of life".

Unfortunately, the magnetic field of the Earth is dying off. While this issue is currently being hotly debated, long term measurements indicate that the overall strength of the field is dropping. This is not good news, in particular since, at the current rate of falloff, it will be reduced to half of its present value in about 1400 years. (156)' ....we have only 2023 years to go. By 3991 A.D. the Earth's magnetic field may have disappeared.' (132) The magnetic field of the Earth is therefore both a circumstances window and a temporal window and as a temporal window, it is rapidly shifting towards closure.

While a minimal or non-existent magnetic field would render the Earth uninhabitable, there is also an upper limit to how much magnetism can be tolerated. In fact, it isn't the magnitude of the magnetism that would provide the upper limit but the source of the magnetism. Magnetism is caused by an electric current. This implies that in the interior of the Earth there is an electric current flowing and that since the magnitude of the field is decreasing, the flow of current must be decreasing. Working backwards, this current would have been greater in the past. The upper limit for habitability would be set by the level of current flow that over-heats the Earth. This is called joule heating. (133) While having a strong magnetic field might be acceptable, the heat that would accompany the flowing current that generates the field would not be acceptable. The heating effect of the current is the upper limitation. In recognition of the current drop-off rate, an unacceptable level of current would have been flowing only about 20,000 years ago. (134) In recognition of the expected lower acceptable limit, 'the window of life' provided by the magnetic field of the Earth is therefore only a few thousands of years wide. By deduction it appears that any candidate-for-life planet in the universe must also have an acceptable magnetic field and it too will be time-wise limited. What is the probability that any such planet will ever be found? Within the Solar System several other objects do have a magnetic field but all of them are too weak to be useful. (This is another reason that Mars will never be habitable.) Never-the-less a magnetic field is a necessity for animal habitability and any far-away planet that is otherwise seemingly appropriate must have a magnetic field within its own particular window of time.

While it is absolutely necessary to have a magnetic field we can see from this discussion that with respect to habitability, it is a fleeting factor. It is temporary. Since cosmic radiation is expectedly present throughout the universe, any planet that seems otherwise habitable must be found with an appropriate magnetic field at the time of interest. While a probability number cannot be identified for this, it is obvious that this will be a demanding criterion to fulfill.

6.7.3   Carbon Dioxide (CO2)

Carbon Dioxide is in the atmosphere of the Earth and it is important to recognize its value with respect to the possibility of life on other planets. There are several reasons why CO2 is absolutely necessary in order to have life – both plant and animal - and any candidate planet must have CO2 in its atmosphere in an amount that is within certain very strict limits. On Earth it also fulfills the function of being a greenhouse gas, without which temperature regulation would be seriously diminished and quite possibly be absent altogether. We need CO2.

All plants are constructed from carbon – the carbon that they extract from the CO2 in the atmosphere. If any plant is to even exist and carry out structural repairs there must be CO2 in the atmosphere. This is true for the grass on the lawn, for the flowers in the garden and for the trees in the forest. All of these things are constructed from carbon taken from the CO2 in the air. While this is true at the present time it was also true long ago regarding the plants which formed the great coal beds which are being mined at the present time.

Coal is carbon which came from plants and the remains of numerous different types of plants have been detected in coal. There is a lot of coal. This means that there is a lot of carbon stored in the great coal storehouse which is really a great carbon storehouse. Some of the plants which have been identified include;' ... tulip tree, magnolia, sequoia, poplar, willow, maple, birch, chestnut, alder, beach, elm, palm, fig, cypress, oak, ... and plum and many other species.' (196) (It will immediately be noted that these are not swamp trees making it clear that coal did not form in swamps.) While this is revealing on its own, the vast quantity of coal opens another avenue of enquiry. The quantity of coal that is still in the ground is exceedingly large and when the amount of carbon that is tied up in this coal is compared to the amount of carbon that is in the current biosphere (i.e. the total of all living things – both plant and animal) it is postulated that there is possibly 50 times as much carbon in the coal as there is in the biosphere. (197) All of this carbon has been trapped away from the carbon cycle and has not been available to re-circulate since it was trapped.

Normally carbon circulates through the biosphere. Animals eat plants and the carbon combines with oxygen in the animal body to produce CO_2 which is expelled back into the atmosphere. Plants can also rot or burn and both of these processes produce CO_2 which is released into the atmosphere. While these are the three main ways that carbon gets into the atmosphere, any process that combines carbon with oxygen and forms CO_2 would also release CO_2 into the atmosphere. In recognition of this, the carbon which formed the coal-bed plants must have come from some source other than the metabolism process of animals because the animals needed the plants in the first place. Neither did it come from the burning or rotting of the plants that are forming the coal because the plant-carbon that formed the coal is still there in the coal. (We must recognize at this time that while the carbon in the coal is demanding of an explanation so is the carbon that is dissolved in the ocean including that which is locked up in other ways in the ocean.) However, it must have come from the air because that is the source of all plant-forming carbon. But how did it get into the air? Where did it come from? What was its source? What process transpired on Earth in order for carbon (i.e. CO_2) to become present in the atmosphere? What process could transpire on a far-away planet to admit CO_2 into its atmosphere?

We understand that an explanation for this dilemma has never been offered. However we will examine three possibilities.

1. The first possibility is that the required CO_2 was formed by a burning process, which used primeval (i.e. virgin) carbon as a source. This burning process introduced the virgin carbon into the atmosphere and into the carbon cycle at just the right rate to enable the trees and other plants, which would later form the coal beds, to grow without producing an over-load and upsetting other life-enabling processes like temperature control. In order to be internally consistent, it must be recognized that a vast amount of virgin carbon was required. In fact exactly the same amount of ancient virgin carbon was required as is presently found in the coal beds (as well as that which has been removed since coal mining began) and in the ocean. (This type of arrangement is recognized as an artificial construct because nature has been conveniently arranged to bring about a result, which isn't otherwise credible.)

2. Is it possible that there could have been enough carbon dioxide in the air at some ancient time to enable the coal-forming plants to develop simply by depleting this CO_2? The amount of carbon dioxide which is in the atmosphere at the present time is more than 380 parts per million (198) which means that 380 out of every million molecules in our atmosphere at the present time are carbon dioxide molecules. If all of the carbon in this carbon dioxide were assembled together to make coal, about $6 \times 10(11)$ metric tons of coal would result. Current estimates of the world's coal reserves are $12 \times 10(12)$ metric tons. (199) The current atmospheric carbon inventory is therefore equivalent to $6/120 \times 100$ or about 4.6% of the world's coal reserves. Therefore, if, prior to formation of the coal bed plants, the amount of atmospheric CO_2 had been about 25 times as great as it is now, the coal bed plants could have grown and depleted that higher level of CO_2 down closer to its present level. (This ignores the vast amount of carbon in the ocean.) Therefore, one possibility to explain the great quantity of carbon required to form the coal-bed plants plus the current inventory of ocean carbon is for an ancient CO_2-rich atmosphere to have been depleted of its CO_2 burden down to an atmosphere with much less CO_2 and in fact just the right amount.

3. Prior to the formation of the coal beds there was an ancient biosphere which was 50 times more extensive than the one we have at the present time. Forests, swamps and meadowlands

were filled with an abundance of all kinds of plants. In addition a greater area of the Earth was involved including the high arctic lands, Antarctica and some areas now below sea level. Then suddenly this ancient biosphere was annihilated and its carbon is now found in coal. This explanation basically shoves the question back because now we must ask where the carbon came from to form this massive assembly of ancient plants. Did they just suddenly appear? Were they created?

In summary, there appear to be three possibilities for the formation of the coal-forming plants.

1. The ancient biosphere was formed from CO2 which was produced by an unknown, virgin-carbon burning process.
2. The ancient biosphere (with all of the coal-bed plants in it) was formed by depleting an even more ancient atmosphere of its CO2 down from a level more than 25 times as high as the present level of 380 ppm (i.e. parts per million).
3. A massive ancient biosphere was created and the plants from it became available to form coal.

However, all of these possibilities come with attachments. If the ancient atmosphere had 25 times as much CO2, the average surface temperature of the world (due to the greenhouse effect) would have been much higher and the trees would not have been able to grow properly because they would have sweat too much. Also, the surface temperature (of the Earth) would have been above the body temperature for most types of animal life as well as too high for the trees themselves. In addition it would have been above the temperature at which seeds can germinate.

While the third possibility is totally unacceptable to many people, with both the first and second, it would have been necessary that none of the plants which grew during the extended times required, were burned, eaten or decayed because this would have released their carbon back into the atmosphere and it would not have been available to contribute to the coal beds. There were no forest fires caused by lightning (which currently strikes the Earth several thousand times every day). Also it was a rot-free forest wherein no significant quantity of material was eaten.

Therefore it can be seen that trying to come up with an explanation for the presence of CO2 in the atmosphere of the Earth is not a straight forward task. However, since CO2 is absolutely necessary for both plant life and animal life to exist, one wonders how an appropriate arrangement could be set up on a far-away planet. The probability of this happening is obviously extremely low and well below any level that could be meaningfully defined.

Carbon dioxide forms a very small portion of our atmosphere but is a major participant in most of life's processes. It is the carbon in carbon dioxide that provides the atomic building blocks so that plants can grow. Trees, as well as grass and all of the plants that are used for food are made from the carbon in carbon dioxide. However there is a limit to the quantity of carbon dioxide in the air that is acceptable. If there was too much, some of the oxygen in the air would be displaced and this would make breathing more difficult for all types of animals. While 'more' enhances the ability of plants to grow, 'too much' inhibits animals from breathing. Too little is also a problem. A reduced amount of carbon dioxide would result in stunted plant growth. Animals could tolerate a modest reduction but plant life would not be vigorous (Animals would likely be much smaller. Herein is a possible explanation for the giant creatures of old.). This makes carbon dioxide a two-sided window and significant deviation from the present level in either direction would not be beneficial.

As mentioned earlier, carbon dioxide is also involved with temperature regulation. The temperature of the Earth must be regulated. In fact, the range of temperature that we can tolerate is really very small. The temperature regulation characteristic of carbon dioxide is manifest through the Greenhouse Gas inventory. Greenhouse Gases have the characteristic of absorbing and re-radiating heat from the surface of the Earth right back to the surface of the Earth. Without this factor operating in our atmosphere, the temperature around the world would deviate much more. It is the combination of the incoming heat from the Sun and the heat absorption and reflection factor of the greenhouse gases that keeps the Earth's surface temperature in the habitable zone. If our Greenhouse Gases were lost, the Earth would not be habitable. On the other hand if the Greenhouse Gas inventory increased too much, the Earth would over-heat and not be habitable for that reason either. In fact, there is serious concern at the present time that our carbon dioxide inventory is increasing too much and that this is resulting in a world-wide increase in temperature. Even a few degrees is anticipated to be a disaster. Carbon dioxide is therefore a 'window of life' and a two-sided window at that. We cannot tolerate too little and neither can we tolerate too much. The over-bearing and seemingly unanswerable question is; how did we magically obtain just the right amount?

In order for any planet in a far-away location to be able to be supportive of both plant and animal carbon-based life, its atmosphere must include carbon dioxide. However trying to identify how this could have been achieved by any natural process has never been explained. The carbon in carbon dioxide is the carbon that is needed to construct plants and from which plants are formed. However there is a very narrow window concerning the amount in the atmosphere that can be tolerated. Too much would cause over-heating. Too little would not enable enough heating. Since carbon dioxide is absolutely required to enable plant growth, if there wasn't any from the beginning, plants could never have started growing. Of course without plants animal life would have been impossible. Further, if the inventory of carbon dioxide had been built up from zero, temperature regulation in a range that was supportive of plant growth would never have happened because the Earth would have been found in a temperature range that was considerably below the freezing point of water and which would have (because of the wide formation of ice and snow) locked-up the Earth's temperature considerably below freezing. (100) This would have been a condition from which it would have been impossible to escape. Clearly having the right amount of carbon dioxide in the atmosphere of either a far-away planet or of the Earth is not a trivial matter but trying to explain how this could ever have been achieved is simply not possible.

6.7.4 Trace Elements

All animal and human bodies require very small quantities of very particular elements. For example the following list shows some of the elements that a human male (age 50-70) requires in his diet. There are others of course. Females require a slightly different assortment. Children also require a slightly different assortment. In fact, all forms of animal life require very small quantities of various elements.

Component	Amount (mg/day)
Calcium	1500
Phosphorus	700

82

Magnesium	420
Iron	8
Zinc	11
Manganese	2.3
Fluoride	4

While this list shows very small quantities of these various items, with even the smallest, several millions of atoms will be involved. It isn't optional whether or not these elements are available, it is absolutely necessary. Some of these items will come from plants that are eaten. Some will come from animals that are eaten. Upon being ingested in a useful form into an animal body these elements will be taken (one atom at a time) to the places in that body where they are required thereby enabling the various body processes to carry on. While we are trying to decide if a far-away planet is habitable it will be necessary to insist that a list similar to the above be available. This supposedly would mean that the creature of interest (i.e. person or animal) must take in the trace elements by eating some plants and/or eating some other creature which has eaten some plants. Now we focus on the plants. The plants would have recovered these elements from the soil. Why different plants take up very particular elements seems rather mysterious but they do. Bananas, for example, take up potassium. In any event the plants take up various elements and directly or indirectly pass these elements along to whatever creature eats the plants. This in turn means that the various elements had to be available in an appropriate form in the soil.

The required soil had to be in place and have within it the elements of interest. The plants could not take up lumps of elements. They would only be able to take up very small assemblies of elements right down at the atomic level or very close to it. How would such small assemblies of atoms of any particular element have gotten into the soil in the first place? Why is it that on Earth certain plants only take up certain elements? Trying to explain where original soil comes from is most difficult to begin with but trying to explain how trace elements would appear in that soil is even more difficult. For example, soil usually forms from plants. Plants grow and die and turn into soil. The same thing happens with most vegetable matter. When it dies (if it is not eaten) it will be converted (by bacteria and oxygen) into soil. While this is a most mysterious process trying to explain why it happens is practically impossible. How, on a far away planet, would this happen? If any plants on that planet were to be useful for animal or human development they must have taken up trace elements (in a useful form) from that far-away planet's soil.

These are very important matters with respect to habitability. In order for a far-away planet to be habitable there must be in place a chain of events that brings very small concentrations of elements from the rock material of that planet (as an aside, why would anyone expect that the rock would include the required trace elements?) to the creatures that will inhabit that environment. What are the chances that a planet with such a process in place will ever be found? This might be called improbable or it might be called impossible. It will never-the-less be a definite requirement in order for life to exist on that planet.

6.7.5　An Atmosphere

The Earth has an atmosphere and it is a very particular atmosphere. The largest portion is nitrogen which, in addition to providing several important functions in life processes, 'waters down' the oxygen so that everything will not be on fire all of the time. (We sadly recall the fire that killed the Astronauts sitting in their ship on the launch pad during the Apollo program. The atmosphere of that ship was 100% oxygen and following that incident the atmosphere was changed to be less hazardous.)

The Greenhouse Effect was discussed above and it was clear from that discussion that an atmosphere, in order to provide temperature control, must include a Greenhouse Effect which of course is provided by very small portions of certain gases in the atmosphere.

If the atmosphere of a far away planet was to support carbon-based life, that atmosphere would need to be very similar to the atmosphere of the Earth otherwise that atmosphere, besides not being supportive of temperature control and various life processes, could actually contaminate the ocean and render it unsuitable for life. Everything must work together to provide life support. The atmosphere would be just as critical as the ocean and the land. While there isn't any way to place a probability on the likelihood of this system necessity happening, it is plain to see that it would be extremely unlikely.

6.7.6 Environmental Stability

In order for any planet to be habitable there must be long-term environmental stability. The Earth no longer enjoys environmental stability and is currently being observed to be warming up. This will be a disaster because of the very narrow range of temperature that is allowable for life to exist. Our temperature requirements are very demanding and the temperature simply cannot be allowed to drift more than a few degrees or life on the Earth will be terminated. Why would we expect anything to be different on a far away planet?

The environmental instability that is now evident on Earth is due to the nature of air with respect to water vapor. Very simply the warmer the air is, the more water vapor it can hold. So if the air is caused to be warmed by any source whatsoever it will be able to hold more water vapor. Unfortunately water vapor is a Greenhouse Gas and more of the Sun's heat will be retained. In turn the additional heat will warm the atmosphere even further causing even more heat to be retained. This is called a viscous cycle with no end in sight that would enable the Earth to remain habitable. Since water vapor is a necessary factor enabling life to exist, one is caused to ask why the Earth has remained habitable for so long? This dilemma is discussed further in 'The non-Myths of the Bible'.

7.0   Finding a Suitable Moon

The search for extra-terrestrial life invariably involves discussion of a planet. Finding a planet within the thermally-habitable zone of a distant star usually causes great excitement - even if the star is a Red Dwarf with no hope whatsoever that any appropriate planet will be found. If a newly-discovered planet is within a star's thermally-habitable zone, there will invariably be some discussion of whether or not the planet might be habitable. The possibility that there will be a nearby moon never comes up. There is justification for this because while planets are barely-discoverable, moons are really not discoverable - at least not yet. However as we have seen from the above discussion an appropriate moon is a necessity in order for a planet to be habitable.

There is no shortage of moons throughout the Solar System. The only planets that do not have some semblance of a moon are Mercury and Venus. Earth has one moon. Mars has two objects which are occasionally referred to as moons but in reality would be more properly referred to as asteroids. Phobos is approximately 8 miles in diameter and Deimos is about 14 miles in diameter. Both of them are non-circular so referring to 'diameter' must be understood to be a term that is taken with a certain amount of licence.

All of the outer planets have moons and it seems that the larger the planet, the more moons there will be. Jupiter, the largest has at least 64 moons. Saturn has about the same and Neptune and Uranus each have several. In the case of Saturn, while there are several objects that are indisputably moons there are several objects within the ring system that might be considered moons and if the rest of the ring material was absent they would likely be referred to as moons. There are plenty of moons in the solar system. Even Pluto has several moons. However the Earth's moon is special. It is special with respect to its size in comparison to the size of the Earth. While several of the outer moons are quite large, the Moon is very large in comparison to Earth. In comparison to Jupiter all of its moons are small. A similar situation exists for the rest of the large outer plants.

The Moon is a necessary factor for the habitability of the Earth. First there is orbital stability and it is understood that the Moon is partially responsible for the orbital stability of the Earth. The Earth's orbit cannot be allowed to wonder from its near-circularity as it must stay very close to the central region of the Sun's habitable zone. Wondering to the fringes of the habitable zone would spell disaster for habitability and humanity with it. The Earth would either chill too much or over-heat. Chilling would spell disaster because the Earth could quite easily become a 'Snowball Earth' as previously mentioned. Once the temperature dropped by a critical small amount, snow and hoar-frost would cover a significant fraction of the surface and there would be no recovery. Things would not be any better in the other direction. A little too much heat and the temperature would continue to rise without any further interference. (i.e. due to positive feedback loops as discussed earlier) This would also be a disaster and this appears to be what is happening to the Earth at the present time. We might conclude in the face of these realities that we can use all of the stability factors that we can find but very tight orbital stability is a minimum necessity for the Earth to remain habitable.

Tides are recognized as beneficial for various forms of wild-life that live adjacent to the ocean. They are also recognized as having a mechanical cleansing action for coastal waterways. Our particular concern however, is habitability and as discussed in various sections above, the tide is a key factor in the establishment and maintenance of the Axial Tilt. The Moon pulls up the tide but if the Moon was smaller the tide that would result would also be smaller. It seems that this would not work because the tide works with the equatorial bulge to establish our necessary Axial Tilt. The Moon, the Tide, and the rotation of the Earth all work together to enable an appropriate Axial Tilt to exist. A smaller Moon and a smaller tide would not work in which case none of this discussion would be taking place.

With respect to the search for a habitable planet in some far-away place it seems that the appropriate planet must also have an appropriate moon. Given the great range in the sizes of the moons in the Solar System, what can be expected for a distant planet? Would it have a large moon or would it have a small moon. Would it have any moon? If the moon in such a situation was not of an appropriate size, the search for life would have to move on. While a probability factor cannot be applied, it is readily recognized that only a very particular arrangement will be satisfactory. When this criterion is added to the criteria recognized so far the probability of finding an overall appropriate situation is vanishingly small.

8.0   Finding a Suitable Universe

Just when we thought that things could not get any worse they did. It is also currently recognized that not only do all of the above-mentioned factors need to line up properly but the universe must have been set up properly in the first place. This is recognized through the study of so-called 'Modern Physics' or more specifically 'Quantum Mechanics' the study of which is too abstract for the average person. In spite of this a review of several of the relevant factors is in order at this time.

First we have the Strong Nuclear Force. It is currently understood that 'a change of as little as 0.5% in the strength of the Strong Nuclear Force would destroy nearly all of the carbon or all of the oxygen in every star and with it the possibility of life as we know it.' (176) A similar thing would happen if the Electric Force was changed. One is prompted to ask why these factors have occurred so particularly? This seems most unlikely but there is more.

Most of the constants that are used in current theories are fine-tuned because if they were altered by very small amounts the universe would, in most cases, be unsuitable for the development of life. (205)

Further is seems that if protons were only 0.2% heavier they would decay into neutrons (which would destroy every atom in the universe). (205) Factors such as these are recognized as a series of 'startling coincidences' because they appear so extremely unlikely.

The conclusive comment with respect to these unlikely factors is that 'Our universe and its laws appear to have a design that is both tailor made to support us and leaves very little room for alteration.' (206)

9.0   Discussion

There are very few stars in the galaxy that can meet the three basic requirements for habitability including a, being in the Galactic Habitable Zone (in fact being in the Galactic Habitable Zone but not within any one of the arms of the Galaxy), b, being hot enough and c, being singular. Even if we start with 300,000,000,000 possibilities these three factors alone bring us down to a few million at the most. To this we must add (for orbital stability) the requirement for a Jupiter at just the right distance. Next we need a Moon of the correct size and at the right distance. Clearly the probability of finding such a combination is exceedingly small. Finally there is the planet itself.

The size-criteria for a potentially-habitable planet has been discussed briefly and it is immediately apparent that very few far-away planets will be able to meet the criteria - even though it seemed reasonably generous. Another factor that was mentioned was the greater attraction that a larger planet would have for whatever space material was in its vicinity. While material is being added to the Earth at the rate of 3000 metric tons per day, (155) a larger planet would supposedly only attract a greater quantity of it. This is not good news for the long-term stability of any planet, including Earth. (Clearly, if one is thinking in terms of billions of years, this problem becomes very significant because an increasingly heavier Earth would pull the Moon continually closer.) All of the factors that make the Earth habitable must have long-term stability. Similarly all of the factors that would make any other planet habitable must also have long-term stability. While a heavier Earth would put a greater pull on the Moon, it would also put a greater pull on all other forms of space material, including asteroids. This would put the Earth at greater risk of being impacted. An inhabited Earth simply could not stand being impacted by even a modest-sized asteroid because earthquakes well above a 9.0 rating would be generated and monster globe-encircling tsunamis would develop and the entire world would be over-run by fast-moving water. This would happen because, compared to the average depth of the ocean (at 2 ½ miles (102)), the land (which has an average elevation above sea level of less than one-fifth that amount (103)) comparatively-speaking barely protrudes above the ocean at the present time and would be easily over-run by any tsunami which was one mile high. In other words, even though some commentators extend the habitability criteria to allow 'earths' up to twice as large as Earth, the habitability of such a place would be much lower than it is on Earth because it would have much greater potential to attract more life-threatening objects and debris from surrounding space.

The habitable zone of the Sun was also discussed. It has frequently been mentioned that the Earth can only tolerate an orbit that keeps it within 5% in either direction of its present distance from the Sun. (178) In other words an orbit that is very close to circular is required. Why is the Earth found right in the middle of the Sun's habitable zone and why does it stay in this location all of the time? Is this coincidental? Or is it just lucky? But more importantly, does it mean that other planets are always required to ensure the orbital stability of a planet of interest? How often is this happening throughout the Galaxy? Do the planets deemed appropriate size-wise have companions which stabilize their orbits to virtual circularity?

Another criterion that planets of interest apparently must have is an associated planetary-vacant region. It seems that other planets are required for orbital stability but none can be too close or there would be no hope whatsoever that orbital stability could be achieved. While Jupiter is required for the orbital stability of the Earth, having Jupiter a little too close would be a disaster. These seem like impossible criteria to meet for any solar system and would obviously limit the possible locations in the universe where a habitable planet might be found to virtually zero.

With an understanding of how the Earth operates we are in a better position to recognize the parameters that a far-away planet must have in order to be habitable. Habitability is not a trivial affair. Even with an ideal planet like the Earth, it is clear that a complete set of favourable and appropriate conditions must be in place before habitability becomes possible and even then several of these conditions are only temporary and several more are easily upset. Is there any realistic possibility that a far-away planet could possess such a coincidental assortment of factors? These factors really do present a complex situation which in no way can be called simple. Could it be replicated elsewhere? It is common to talk about probability in situations like this but in this case there are no reference points from which a probability factor could be derived. We could resort to simple reasoning and ask the question; based on the complexity of the situation on Earth, what is the possibility that a similar situation exists elsewhere? Such a possibility is obviously someplace between very slim and non-existent but from a straight-forward approach how the situation that the Earth enjoys could ever have developed is really impossible to explain. Further, even though the situation on Earth is close to ideal it is a transient and will not remain in place forever.(i.e. many of the 'windows of life' have only a short-term existence.) We recall that the Earth's axial tilt is dependent on the rotational speed of the Earth so as the speed decreases the axial tilt will be in jeopardy. Further, how did the axial tilt develop? With the Earth rotating much faster in the past it would have been sitting up straight. How could it have been tipped over to an angle that is ideal for temperature control and hence for habitability? How did the Earth acquire its ideal axial tilt? While this situation is totally unexplainable for the Earth (about which we know a considerable amount) what could we possibly expect from a far away planet?

10.0   Conclusion

With respect to providing the right amount of heat for a potentially-habitable planet, the number of candidate stars available is really very low. When the constraints that a planet and its associated moon must have are included, a situation results that is extremely improbable even before temperature control and life-support factors are added. When it is recognized that the entire setup would only last for a few (thousand) years at the most, the situation is truly hopeless. In reality, the circumstances required for life-support go far beyond those mentioned herein. Therefore suggesting that the universe is teeming with life is not realistic. The number of coincident factors required is so incredibly high that it makes the probability of extraterrestrial life virtually zero and we are justified in concluding that there is no life elsewhere in the universe. In fact, what would be required would be for another solar system just like ours complete with a sun just like ours, a jupiter just like ours, (complete with a set of other planets to stabilize its orbit) a moon just like ours and a planet like the Earth complete with a warm, semi-molten interior, a large ocean and an atmosphere that had an active Greenhouse Effect. From the viewpoint of habitability (recognizing the necessity for very tight temperature control) nothing less would do. It must be located in The Galactic Habitable Zone outside of the arms of the Galaxy. With the recognition that there really are only a modest number of appropriate stars to begin with, it would be bizarre to think that such a duplicate solar system actually exists. Therefore it is appropriate to conclude that;

Extraterrestrial Life is non-existent and, in fact, Impossible!

Appendix One - Bibliography

Reference	Abbreviation
Funk & Wagnalls New Encyclopedia, by Funk & Wagnalls Inc., United States of America	F & W
Creation Matters, Creation Research Society, P O Box 8263, St. Joseph MO, 64508-8263 USA	CM
Creation Research Society Quarterly, 6801 N, Hwy 89, chino Valley AZ 86323	CRSQ
Apocalypse When? Cosmic Catastrophe and the Fate of the Universe by Frank Close, William Morrow and Company Inc., New York	Cosmic
The Sea Around Us by Rachel Carson, Oxford University Press Inc. 2003, 198 Madison Ave New York, NY	The Sea
Critique of Radiometric Dating by Harold s. Slusher Institute for Creation Research, San Diego, CA 92116	Critique
The European Discovery of America by Samuel Eliot Morison, Oxford University press, New York	European
Field Notes from a Catastrophe by Elizabeth Kolbert Bloomsbury, New York	Field
Silent Snow, The Slow Poisoning of the Arctic by Marla Cone, Grove Press, New York	Silent
Climate and the Oceans, by Geoffrey K. Vallis, Princeton University Press, 41 William Strre, Princeton, New Jersey 08540	Climate
College Physics by Weber, White and Manning, McGraw-Hill Book Company, New York NY	College
Comets and Asteroids and Future Cosmological Catastrophes compiled by Glen W. Chapman, www.2s2.com/chapmanresearch	Comets

Reference	Abbreviation
Design and Origins in Astronomy, By George Mulfinger, Jr., Creation Research Society Books	Design
The Moon, Its Creation, Form and Significance by John C. Whitcomb and Donald B. DeYoung BMH Books, Winona Lake, Indiana	The Moon
Funk & Wagnalls New Encyclopedia Dun & Bradstreet Corporation	F&W
A Short History of Nearly Everything by Bill Bryson Anchor Canada	Short
The living Cosmos by Chris Impey Random House, New York	Living Cosmos
National Geographic, Official Journal of the National Geographic Society	Nat Geo
Canadian Geographic The Royal Canadian Geographic Society	Can Geo
The Toronto Star	Tor Star
Scientific American Scientific American Inc. 415 Madison Avenue, New York N. Y. 10017	Sc Am
How It Works Imagine-Publishing.co.uk	HIW
Discover Magazine Kalmbach Publishing Co.	Discover
The Oceans by Ellen J. Prager McGraw-Hill New York	The Oceans

Reference	Abbreviation
Climate Wars by Gwynne Dyer Random House Canada	Climate Wars
Earth in Upheaval by Immanuel Velikovsky Out of print	E in U
Book of the Cosmos David H. Levy, Editor Scientific American	Cosmos
Weather, a Visual Guide by Bruce Buckley et al Firefly Books	Weather
Chemistry by Quagliano Prentice Hall Inc, Eaglewood Cliffs, New Jersey	Chem
CRC Handbook of Chemistry and Physics	CRC
The Grand Design by Stephen Hawking Bantam Books, The random House publishing Company, random House, New York	Grand

Appendix Two   References

Number	Source	Topic
100	Climate p 671	Snowball Earth
101	Comets p 51	giant comet into the Sun
102	F&W V8 p 423	average ocean depth
103	F&W V8 p 423	average land elevation
104	Short p 247	hab. zone + 5% or - 15%
105	CRC p F147	Earth orbit deviation +or- 2%
106	The Moon p 86	200,000 impact marks on the Moon
107	Starchild meteoroids	Earth's mass is increasing continually
108	Living Cosmos p 291	Intelligent life would annihilate itself
109	Nat Geo Dec 89 p 91	Universe very queer
110	wiki multiple stars	examples of triples
111	CM j/f 2005 p 1	90% of stars are multiples
112	Sc Am April 1977 p 98	nearby multiples
113	Tor Star April 1999 A 3	planets would be flung out
114	Cosmos p 126	75% binary
115	Cosmos p 126	Sun in a minority of singles
116	Sc Am April 1977 p 98	cannot detect all companions
117	www.bbc.com.news	planet orbiting Proxima Centura
118	Tor Star Feb 23 2017	dwarf stars produce planets
119	Tor Star Feb 23 2017	Trappist-1 has 6 stars in habitable zone
120	www.icr.org	only 5% of all stars are bright enough
121	Wiki Gliese 581g	ice caps on night side
122	wiki Gliese 581	Gliese 581
12	3icr.org.habitable zone	galactic habitable zone
124	Inco Creighton Mine	warm nickel mine
125	HIW Issue 36 p 97	temperature increases with depth
126	HIW Issue 36 p 57	Kola borehole temperature
127	Discover J/A 2018 p 56	Antarctic borehole
128	Field p 106	CO2 driving factor

Number	Source	Topic
158	European p 595	current in Hudson Straight
159	Climate p 165	average surface temperature
160	Climate p 21	snowball Earth
161	Climate p 2	ocean size and depth
162	Evolution news.org	GJ581d
163	Climate p 21	drop in temp. would lose water vapour
164	Climate p 1	planet influence on Earth's orbit
165	Cosmos p 191	Jupiter influence on Earth's orbit
166	Climate p 165	average temp 15C
167	Field p 30	albedo of snow
168	CRSQ v 15 p 153	albedo of land
169	Climate p 105	ocean heat storage
170	Silent p 128	Faroe Islands temperature
171	Climate p 119	Gulf Stream
172	Grand p 150	circular orbit most fortunate
173	Climate p 107	Climate of New York and San Francisco
174	Cosmic Pursuit spring 1985	planet size Earth 0.83 to 1.2
175	Tor Star Feb 23, 2017	starlight dips when planet crosses
176	Grand p 159	catastrophic if strong nuclear force is changed
177	Cosmic Pursuit spring 1985	Mars ice-age if water present
178	CRSQ v 32 p 76	habitable zone + or - 5%
179	Comets p 79, Design p 127	Sun becoming smaller and dimmer
180	Climate p 21	recovery from freeze-up is impossible
181	Cosmos p 294	Earth has a Goldilock's orbit
182	Short p 248	molten interior beneficial
183	Weather p 32	Hadley Cell circulation
184	The Oceans p 198	Chinese Tidal Bore
185	Silent p 128	Faroe Islands temp.
186	Silent p 249	Moon stabilizes axial tilt
187	E in U p 19	frozen trees up north

Number	Source	Topic
188	Oxford p 624, The Moon p 145	Moon librates
189	CRSQ v 37 p 185	1178 asteroid into Moon
190	Cosmos p 319	water must have a neutral pH
191	Chem 468	pH > 7 = basic < 7 = acidic
192	Cosmos p 318	stable Jupiter orbit needed
193	Nat Geo July 2014 p 42	habitable zone includes Mars
194	Cosmos p 227	Mars weather
195	nasa.gov.ames.kepler	kepler186f in habitable zone
196	E in U p 202	trees in coal
197	CSRQ v 20 p 218	50x carbon in coal
198	Field Notes p 43	CO_2 in air 380 ppm
199	CSRQ v 20 p 218	coal reserves
200	CSRQ v 32 p 26	habitable zone 5% closer
201	CRSQ v 29 p 90	not enough light up north
202	Climate Wars p 95	CO_2 & Methane from Russian bogs
203	astrobiology.nasa.gov	galactic habitable zone
204	Wiki Gliese 581c	planet tidally locked
205	Grand p 160	constants cannot vary
206	Grand p 162	universe tailor-made for us

Appendix Three Index

www.ingramcontent.com/pod-product-compliance
Lightning Source LLC
Chambersburg PA
CBHW061617210326
41520CB00041B/7485